NLM dup.

Fundamentals of Nuclear Medicine Dosimetry

Michael G. Stabin, PhD
Associate Professor, Department of Radiology/Radiological Sciences, Vanderbilt University, Nashville, Tennessee, USA

Fundamentals of Nuclear Medicine Dosimetry

9/08

Michael G. Stabin, PhD
Associate Professor
Department of Radiology/
Radiological Sciences
Vanderbilt University
Nashville, TN, USA

ISBN: 978-0-387-74578-7 e-ISBN: 978-0-387-74579-4
DOI: 10.1007/978-0-387-74579-4

Library of Congress Control Number: 2007937210

Printed on acid-free paper

9 8 7 6 5 4 3 2 1

springer.com

Preface

If you reveal your secrets to the wind, you should not blame the wind for revealing them to the trees.

—From *The Prophet* by Kahlil Gibran

This was a very satisfying book to write, but also perhaps a very dangerous one. After 15 years of building a base of knowledge in radiopharmaceutical dosimetry (under the expert guidance of Evelyn Watson at the Oak Ridge Dosimetry Center), 8 years of teaching radiation protection and dosimetry at the university level (in northeastern Brazil and at Vanderbilt University in the United States), and authoring a number of papers and book chapters on the subject, it was rewarding to write down in a single, organized text nearly all that I have learned on this topic.

My deepest desire is that this text will be found useful by members of the nuclear medicine and radiation protection community in formalizing and facilitating such calculations. After learning a technique for performing dose calculations that must be done often, I have often found ways to automate such work in software tools and to make such tools available to others in order to help them shortcut their efforts. One of the greatest rewards I have received in my work has been to see a few of the tools and ideas that I have developed become useful to others in their routine work.

In this text, I reveal practically all of my methods and secrets for practical internal dose calculations. Some enjoy keeping their methods and ideas quite private, to protect their personal interests, and such persons may consider me to be a fool for sharing all of mine so openly. I have found, however, that the open sharing of information in the scientific community leads to heightened understanding, thriving collaboration between like-minded investigators, the spawning of new ideas, and the growth and maturation of new investigators whose contributions end up benefiting us all. I trust that my efforts here will provide some small impetus toward these ends and that readers will find the information in these pages to be clear, helpful, and frequently referenced for practical use.

Michael G. Stabin, PhD

Contents

1 Uses of Dosimetry Information in Nuclear Medicine

Diagnostic Versus Therapeutic Applications

Since the earliest days after the discovery of radiation in 1895 by Wilhelm Conrad Roentgen, it has been known that exposure to ionizing radiation can be harmful to humans. In any use of ionizing radiation, one must prevent or minimize the risks of the use of the radiation while allowing its beneficial applications. As we will discuss later in this book (Chapter 6), current research is challenging the paradigm that the quantity *absorbed dose* is the best to use in predicting biological effects. There are clearly complications that need to be considered in assessing the response of all biological systems to all kinds of radiation. Nonetheless, the quantity *absorbed dose*, which gives the energy of ionizing radiation absorbed per unit mass of tissue (or any material for that matter), is usually indicative of the probability of a deleterious biological effect, and it is the quantity that will be studied most in this text.

The history of the use of radioactive materials as biological tracers dates to Georg de Hevesy and colleagues, who, in 1924, performed radiotracer studies of the kinetics of lead-210 (^{210}Pb) and bismuth-210 (^{210}Bi) in animals. Soon thereafter in 1925, Herrman Blumgart and Otto Yens evaluated blood flow rates in humans using bismuth-214.

Iodine-131 and cobalt-60 were discovered by John Livingood and Glenn Seaborg, and Emilio Segre and Glenn Seaborg discovered technetium-99m (^{99m}Tc) in 1938. Iodine-131 (^{131}I) and ^{99m}Tc are the predominant radionuclides currently in diagnostic and therapeutic nuclear medicine studies. The ^{99m}Tc generator was developed in 1957 by W.D. Tucker and colleagues at the Brookhaven National Laboratory.[1]

When X-rays were first discovered by Roentgen, the idea that physicians could see internal structures of the body without using a scalpel was one of the most exciting moments in medicine. Similarly, the idea of using a radioactive tracer inside of the body to transmit signals to detectors outside of the body to investigate the movements of materials in the body and thus discern physiologic, as opposed to only anatomic, information was exciting and revolutionary. The first application of nuclear medicine is *diagnostic*; that is, studying structures and processes to diagnose diseases and guide medical response to potential human health issues. The majority of day-to-day practice in nuclear medicine continues to involve diagnostic procedures, but radiopharmaceuticals used in nuclear medicine may also be applied in *therapeutic* applications; that is, administering higher levels of activity with the intent of exploiting the ability of radiation to destroy deleterious tissues in the body (cancer, inflamed joints, and other applications).

In the early part of the 20th century, shortly after the discovery of radiation and radioactivity, radiation sources were used in a number of ill-advised experiments with medical applications and consumer products. Dr. Paul Frame, in a historical review of the use of such applications, notes that popular locations in the United States and elsewhere attracted visitors who could bathe in springs of radioactive water or inhale radioactivity-laden air.[2] Some of these sites are still in operation, surprisingly. Routine intentional exposures to radiation occurred, based on the belief that radiation could cure "various forms of gout and rheumatism, neuralgia, metallic or malarial poisoning, chronic Bright's disease, gastric dyspepsia, chronic diarrhea, chronic skin lesions . . . insanity, old age" and create "a splendid youthful joyous life."[2] Dr. Frame notes that "Professor Bertram Boltwood of Yale explained the scientific basis for the cures in the following way: The radioactivity was "carrying electrical energy into the depths of the body and there subjecting the juices, protoplasm, and nuclei of the cells to an immediate bombardment by explosions of electrical atoms," and that it stimulated "cell activity, arousing all secretory and excretory organs . . . causing the system to throw off waste products," and that it was "an agent for the destruction of bacteria."[2] A series of consumer products arose that allowed people to drink radioactive water on a regular basis in their own homes without having to travel to a spa or mine many hours away. High rates of thyroid carcinoma were observed in the 1940s and 1950s in infants treated shortly after birth for thymus enlargement. In a similar time period, large doses of radiation were given to the spines of subjects suffering from ankylosing spondylitis; the treatment was effective, but it was associated with a high rate of induction of leukemia. Radiologists and radiotherapists operating in the early years of radiation medicine suffered high rates of leukemia and pernicious anemia. A particularly tragic episode in the history of the use of radiation and in the history of industrialism was the acute and chronic damage done to the radium dial painters.[3] Radium was used in luminous paints in the early 1900s, and some factory workers (mainly women) ingested large amounts of radium-226 (^{226}Ra) during the painting of the luminous dials for watches. They soon demonstrated high rates of bone cancer and even spontaneous fractures in their jaws and spines from cumulative radiation injury.

In the early years, ^{226}Ra was the principal radionuclide used in radiation therapy in which high-activity sources were placed on or near tumors to attempt to eradicate them (*brachytherapy*; with *brachy* coming from a Greek word meaning "close to"). Modern external radiation therapy still employs a number of brachytherapy techniques involving different radionuclides and radiation-producing machines that deliver high doses of radiation to malignant tissues while minimizing dose to healthy body tissues. In nuclear medicine therapy, the goal is to administer compounds systemically that will preferentially concentrate in tumors and deliver a high dose to these tissues while hopefully having lower concentrations and faster clearance rates from other tissues. An overview of the current practice of nuclear medicine therapy is given in Chapter 5.

A radiation dose analysis is fundamental to the use of either diagnostic or therapeutic radiopharmaceuticals. For diagnostic compounds, the U.S. Food and Drug Administration (FDA) studies a number of safety issues during the drug approval process, and internal dosimetry* is one issue of high importance (see Chapter 7). Radiation dose estimates are not often of direct interest in day-to-day practice in the clinic, but they are often referred to when comparing advantages and disadvantages of possible competing drug products, by radioactive drug research committees (RDRCs) in evaluating safety concerns in research protocols, and in other situations. In therapeutic applications, the physician should perform a patient-specific evaluation of radiation doses to tumors and normal tissues and design a treatment protocol that maximizes the dose to tumor while maintaining doses to healthy tissues at acceptable levels (i.e., below thresholds for direct deleterious effects), as is always done in external radiation therapy treatment planning. Unfortunately, this is not routinely practiced in most clinics at present, and patients are generally all treated with the same or similar protocols without regard to their specific biokinetic characteristics.

**Dose assessment* is actually the correct, formal name for this process. In day-to-day use, however, the terms external and internal *dosimetry* are used. This is the classic, historical usage since the Manhattan Project in the 1940s. The term *dosimetry* contains the suffix *metry*, which relates to metrology, which implies the measurement of physical quantities. Much of external dose assessment does have to do with measurements, so the term *dosimetry* is mostly accurate, although some assessment is done with theoretical models. The science of *internal* dose assessment is almost entirely founded in theoretical calculations and models, with no measurements being involved. This field is, however, most often referred to as *internal* dosimetry, to be a correlate to *external* dosimetry.

Model-Based Versus Patient-Based Approaches

As noted above, internal dose estimates are performed via calculations, not measurements. Usually, they are based on standardized models of the human body and often on standardized models of radiopharmaceutical behavior in the body as well (see Chapter 3). This approach results in a calculation that is easily traceable and reproducible. One of the important aspects of model-based internal dose calculations is that the output (calculated dose estimates) is only as good as the input (assumptions and models employed). With very good data, one can obtain good dose estimates, but one must always remember that what has been calculated is dose to a *model*, not dose to a *person* (patient, research subject, etc.). In diagnostic applications, this is generally acceptable. All of the input data has some associated uncertainty, and the calculated results reflect both this inherent uncertainty in the data as well as uncertainties related to the application of standardized models of the body to a variety of patients who vary substantially in size, age, and other physical characteristics. When the radiation doses are low, this kind of uncertainty is tolerable because, if the calculated answers are incorrect by tens of percent or even factors of 2 or more, the consequences for the patient are small or nonexistent (depending on what you believe about radiation risk models at low doses and dose rates; see Chapter 6). In therapeutic applications, however, the tolerance for uncertainty needs to be lower, as the doses are higher, and the chances of reaching or exceeding an organ's threshold for expressing radiation damage are real.

The use of model-based dosimetry for therapy with radiopharmaceuticals must be abandoned and replaced with a patient-specific modeling effort that considers both the unique anatomic and physiologic characteristics of the patient, as has been done in external beam radiotherapy for decades. It is true that such attention to detail for the patient's benefit requires significantly more effort in data gathering and dosimetric modeling, but the effort is worthwhile in providing the patient with a better quality of care and expectancy for a positive outcome from the therapy. Radiopharmaceutical therapy must advance beyond its roots in pharmaceutical therapy dosage (e.g., using a quantity of drug per unit measure of patient size) and become more like radiotherapy dose delivery (e.g., employing understandings of energy deposited per unit mass of tissue).

Practical Dosimetry: Balancing Benefits and Risks

As we study the details of dose calculations, we will develop carefully the minimum requirements for performing an adequate calculation of radiation dose. Natural progress will ensure that more and more technology can be brought to bear in analyzing the problem. With more and more data constantly gathered, the quality of the results and approaches will improve. Data gathering for dosimetry requires that the subject spend perhaps 10 to 40 minutes in a fixed geometry in a detector counting system, which involves some discomfort and difficulty. The data analysis requires the time and attention of skilled professionals. There is obviously a balance that needs to be struck between an excellent analysis and logistical concerns. After the absolute minimum of data is obtained, additional data may be taken as is possible, given the concurrence of the physician and patient and availability of the counting systems. When only fixed dosages of radiopharmaceuticals are administered to all patients with no study whatsoever of the radiation doses received, it is impossible to optimize individual subjects' therapies or to advance our understanding of dose-effect relationships and how to provide the best treatments for our patients.

Clinical Utility: Interface of Patient, Physician, and Physicist

The patient-physician relationship is a special one, involving trust, weighing and balancing of very significant decisions, exposure of highly personal information, and usually significant expenses. Decisions about medical procedures and follow-up, use of medications, planning, and lifestyle are very personal and often difficult. The physicist, who provides just one piece of information that the physician must weigh and convey to his or her patients, plays a very peripheral but nonetheless pivotal role in the case of therapeutic use of radiation. In diagnostic applications, the physicist is separate from the process and just provides dose calculations that are used by regulatory and other bodies to make very broad recommendations about the general use of radiopharmaceuticals in clinical practice and research. Only rarely does a flawed diagnostic study (e.g., a misadministration) necessitate the physicist's attention to dosimetry in a particular patient's situation. These three individuals must work closely, however, in this ongoing process to provide the highest quality medical care possible in every circumstance. Ultimately, the patient (or research subject) makes the final decisions about the progress of the medical care and must be given high-quality information, clearly and unambiguously communicated by the physician and/or physicist. It is important that radiopharmaceutical therapy begin to involve the physicist more than it has in the past, as is modeled in external beam radiotherapy.

References

1. Stabin M. Nuclear medicine dosimetry. Phys Med Biol 51: R187–R202, 2006.
2. Frame PW. Radioactive curative devices and spas. Oak Ridger Newspaper 5 November, ID-3D, 1989.
3. Mullner R. Deadly Glow. The Radium Dial Worker Tragedy. American Public Health Association, Washington DC, 1989.

2
Fundamental Concepts: Calculating Radiation Dose

Dosimetry Quantities and Units

Quantification of the amount of radiation received by a potentially radiosensitive site is essential to the characterization of the possible risks of the exposure. The principal quantity used to identify and measure the amount of radiation received is the *absorbed dose*, sometimes called just *dose*. The word *dose* has a number of meanings in its general use.

As a noun, these are the definitions:

1. An amount of some agent applied for a medical purpose:
   (a) A specified quantity of a therapeutic agent, such as a drug or medicine, prescribed to be taken at one time or at stated intervals.
   (b) The amount of radiation administered as therapy to a given site.
2. An ingredient added, especially to wine, to impart flavor or strength.
3. An amount, especially of something unpleasant, to which one is subjected: *a dose of hard luck*.
4. *Slang:* A venereal infection.

As a verb, these are the definitions:

1. To give (someone) a dose, as of medicine.
2. To give or prescribe (medicine) in specified amounts.

In this text, we are interested in the quantity alluded to in part 1(b) above and will very specifically define it. This little diversion was entertained to point out that when one uses the term *dose* in a medical setting, it is not uncommon for the understanding of that term to vary. Many times, physicians refer to the dose of a radiopharmaceutical given to a patient, meaning the *amount of activity given to the subject* (MBq or mCi, for example), not the radiation dose (rad or Gy) received by the tissues of the patient's body. This is a sometimes unfortunate but very understandable mixing of terms, as physicians administer doses of medicine more often than dosimetrists calculate doses of radiation for medical subjects. One must simply be aware of this possible confusion of terms and be sure that the right quantities are employed in the right circumstances. One solution is to use the term *dosage* to refer to the quantity of an administered pharmaceutical and reserve the term *dose* for quantification of radiation dose (i.e., energy/mass).

The first quantity that is of interest to our text is *absorbed dose*. Absorbed dose is the *energy absorbed per unit mass of any material*. Absorbed dose (D) is defined as:

$$D = \frac{d\varepsilon}{dm}$$

where $d\varepsilon$ is the mean energy imparted by ionizing radiation to matter in a volume element of mass dm. The *units* of absorbed dose are energy/mass of any material. One may use, for example, erg/g, J/kg, or others. *Special units* are also defined for absorbed dose:

1 rad = 100 erg/g
1 gray (Gy) = 1 J/kg
1 Gy = 100 rad

The word *rad* was originally an acronym meaning "radiation absorbed dose." The rad is being replaced by the SI unit value, the gray (Gy), which is equal to 100 rad. Note that *rad* and *gray* are collective quantities: one does not need to place an "s" after them to indicate more than one.

As will be shown in Chapter 6, many biological effects of radiation can be related to an amount of absorbed dose.

At very low doses, no effects may be observed. After the dose passes a particular threshold, some effects may be observed and will generally become more severe as more dose is received. However, when different experiments are performed in certain biological systems using perhaps different kinds of radiation or measuring different biological end points, different amounts of absorbed dose may be needed to observe a particular effect. This is particularly true for high linear energy transfer (LET) radiations like alpha particles and fast protons.

The other important quantity traditionally defined to account for these differences is the *equivalent dose*. Equivalent dose is the absorbed dose modified by a factor accounting for the effectiveness of the radiation in producing biological damage. Equivalent dose ($H_{\mathrm{T,R}}$) is defined as:

$$H_{\mathrm{T,R}} = w_{\mathrm{R}}\, D_{\mathrm{T,R}}$$

where $D_{\mathrm{T,R}}$ is the dose delivered by radiation type R averaged over a tissue or organ T, and w_{R} is the radiation weighting factor for radiation type R. The factor w_{R} is really dimensionless; the fundamental units of equivalent dose are the same as those for absorbed dose. Operationally, however, we distinguish using the special units:

H (rem) $= D$ (rad) $\times w_{\mathrm{R}}$
H (Sv) $= D$ (Gy) $\times w_{\mathrm{R}}$
1 Sv (sievert) = 100 rem

Note that, like rad and gray, rem and sievert are collective terms; one need not speak of "rems" and "sieverts," although this may sometimes be heard in informal speech and even observed in some publications.*

The recommended values of the radiation weighting factor have varied somewhat over the years as evidence from

*Also note that units that incorporate a person's name (Roentgen, Gray, Sievert) are given in lowercase when spelled out completely but with the first letter capitalized when given as the unit abbreviation (e.g., sievert and Sv).

biological experiments has changed. The current values recommended by the International Commission on Radiological Protection (ICRP)[1] are given in Table 2.1.

So, for photons and electrons, numerical values of the absorbed dose in gray and equivalent dose in sievert are equal. For alpha particles and other particles, the equivalent dose is a multiple of the absorbed dose.

To estimate absorbed dose for all significant tissues, one must determine for each tissue the quantity of energy absorbed per unit mass. This yields the quantity absorbed dose, if expressed in proper units, and can be extended to calculation of equivalent dose if desired. What quantities are then needed to calculate the two key parameters—energy and mass? To concretely estimate absorbed dose, we must assign numerical values to all of the quantities involved in the energy and mass terms. To study this, we can imagine an object that has a uniform distribution of radioactive material throughout. Depending on the identity of the radionuclide, particles or rays of characteristic energy and abundance will be given off at a rate dependent upon the amount of activity

TABLE 2.1. Radiation weighting factors recommended by the ICRP.

Type of radiation	w_R
Photons, all energies	1
Electrons and muons, all energies (except Auger electrons in emitters bound to DNA)	1
Neutrons, energy:	
$<$10 keV	5
10 keV to 100 keV	10
$>$100 keV to 2 MeV	20
$>$2 MeV to 20 MeV	10
$>$20 MeV	5
Protons, other than recoil protons, $E>2$ MeV	5
Alpha particles, fission fragments, heavy nuclei	20

Source: Reproduced with permission from International Commission on Radiological Protection. 1990 Recommendations of the International Commission on Radiological Protection. ICRP Publication 60. Pergamon Press, New York, 1991.

present. This object must have some mass. Once we know the identity of the radionuclide of interest to our calculation, we can find from many paper or electronic resources values that provide the energies per decay (and number of emissions per decay) for it. We must specify the amount of activity in the object, and we must know the mass of the target region. One other factor that we will need is the *fraction of energy released in the object* that is absorbed within the object. This quantity is most often called the *absorbed fraction* and is often represented by the symbol ϕ. For photons (gamma rays and X-rays), some of the emitted energy will escape objects of the size and composition of interest to internal dosimetry (mostly soft tissue organs having diameters of the order centimeters). For electrons and beta particles, most energy is usually considered to be absorbed, so we *usually assume that the absorbed fraction is 1.0.* Electrons, beta particles, and the like are usually grouped into a class of radiation referred to as *nonpenetrating emissions*, whereas X-rays and gamma rays are called *penetrating radiation*. This is simply an *operational definition* used in internal dosimetry. Certainly, many beta particles may penetrate materials like paper and Mylar and even penetrate the outer layers of the skin and give a radiation dose to sensitive cells in the body.

We can now develop a generic equation for the absorbed dose rate in our object as:

$$\dot{D} = \frac{kA \sum_i y_i E_i \phi_i}{m}$$

where $\dot{D}$ is absorbed dose rate (rad/h or Gy/s), A is activity (μCi or MBq), y is number of radiations with energy E emitted per nuclear transition, E is energy per radiation (MeV), ϕ is fraction of energy emitted that is absorbed in the target, m is mass of target region (g or kg), and k is some proportionality constant (rad-g/μCi-h-MeV or Gy-kg/MBq-s-MeV).

It is *vital* that the proportionality constant be properly calculated and applied. The results of our calculation will be incorrect (perhaps dangerously so!) unless the units within are consistent and they correctly express the quantity desired. The application of radiation weighting factors to this equation to calculate the dose equivalent rate is a trivial matter; for most of this chapter, we will consider only absorbed doses for discussion purposes.

The investigator is not usually interested only in the absorbed dose rate; more likely, an estimate of total absorbed dose from an administration is desired. In the above equation, the quantity activity (nuclear transformations per unit time) causes the outcome of the equation to have a time dependence. To calculate the *cumulative* dose, the time integral of the dose equation must be calculated. In most cases, the only term that has a time dependence is *activity*, so the integral is just the product of all of the factors in the above equation and the *integral of the time-activity curve.*

Regardless of the shape of the time-activity curve, the integral of the curve, however obtained, *will have units of the number of nuclear transitions* (activity, which is transitions per unit time, multiplied by time) (Fig. 2.1). Therefore, the equation for cumulative dose would be:

$$D = \frac{k\tilde{A}\sum_i y_i E_i \phi_i}{m}$$

where D is the absorbed dose (rad or Gy) and $\tilde{A}$ is the number of nuclear transitions, or *cumulated activity* (perhaps given

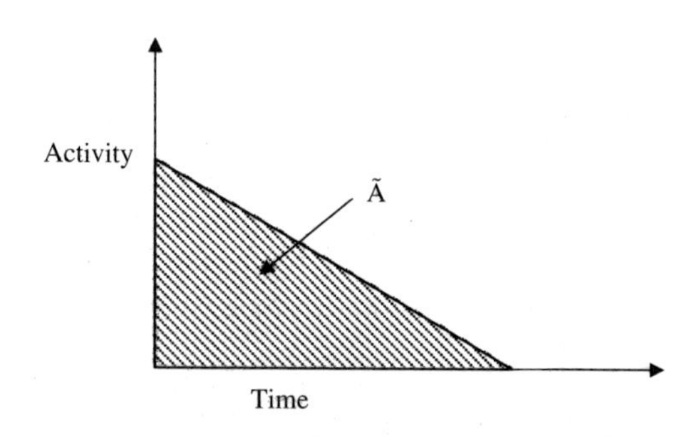

FIGURE 2.1. Generalized internal source time/activity curve.

as μCi-h or MBq-s). The numerical value of k reflects the units chosen for the other terms in the equation. Details and examples of the use of these equations will be developed later in this chapter.

Effective Half-Time

Radioactive materials decay according to first-order kinetics; that is, a certain fraction of the remaining activity is removed during a specific time interval:

$$\frac{dN}{dt} = -\lambda N$$

The well-known solution to this equation is:

$$N(t) = N_0 e^{-\lambda t} \qquad A(t) = A_0 e^{-\lambda t}$$

Here, N is the number of atoms, N_0 is the initial number of atoms, A is the amount of activity at any time t, and A_0 is the initial activity. Fortunately, many materials are also cleared from the body or certain organs by first-order processes. We can thus develop an equation for the reduction in the amount of a *nonradioactive* substance:

$$X(t) = X_0 e^{-\lambda_b t}$$

where $X(t)$ is the amount of the substance at time t, X_0 is the initial amount of substance X, λ_b is the biological disappearance constant ($= 0.693/T_b$), and T_b is the biological halftime for removal.

A biological halftime for removal is exactly analogous to a radioactive (or physical) halflife; that is, it is the time in which half of the remaining material is removed, but here only by biological processes.

If we now consider a certain amount of *radioactive* material in the body that is being cleared from the body by a first-order process, two first-order processes will be involved in removing the activity from the body: radioactive decay and biological disappearance. Because these decay constants are essentially probabilities of removal per unit time, the disappearance constants for the two processes can be *added* to give an *effective disappearance constant*:

$$\lambda_e = \lambda_b + \lambda_p$$

where λ_e is effective disappearance constant, λ_p is radioactive (physical) decay constant, and λ_b is biological disappearance constant.

We can also define an *effective halftime* equal to $0.693/\lambda_e$, which is the time for half of the activity to be removed from the body or organ, by both physical decay and biological removal. It can be shown easily that the effective halftime is related to the other biological and physical halftimes by the following relationship:

$$T_e = \frac{T_b \times T_p}{T_b + T_p}$$

For materials that follow this relationship, the integral of the timeactivity curve may be easily evaluated:

$$\tilde{A} = \int_0^\infty A(t)\,dt = \int_0^\infty (fA_0)e^{-\lambda_e t}\,dt = \frac{(fA_0)}{\lambda_e} = 1.443fA_0T_e$$

where A_0 is the activity administered, f is the fraction that goes to a given region (sometimes called an *uptake fraction*), and $(f \times A_0)$ is the initial amount of activity in the region. So, effective half-time is a critical parameter in the determination of cumulated activity and cumulative dose.

Note that the effective half-time for a compound will always be *less than or equal to the shorter of either the biological or radiological half-time*. As two processes are contributing to the removal of the element, the action of the two together must act faster than either acting alone. Note also that to solve the equation for effective half-time, the units for the biological and physical half-times must be the same.

Examples

$$T_b = 7 \text{ days } T_p = 20 \text{ days} \qquad T_{eff} = \frac{20 \times 7}{20 + 7} = 5.19 \text{ days}$$

$$T_b = 7 \text{ days } T_p = 7 \text{ days} \qquad T_{eff} = \frac{7 \times 7}{7 + 7} = 3.5 \text{ days}$$

Note: This is not a coincidence. Every time that the biological and physical half-times are the same, the effective half-time is exactly half of either value, because the value is $(x \times x)/2x = x/2$.

$$T_b = 7 \text{ days } T_p = 100 \text{ days} \qquad T_{eff} = \frac{100 \times 7}{100 + 7} = 6.54 \text{ days}$$

$$T_b = 7 \text{ days } T_p = 10^9 \text{ days} \qquad T_{eff} = \frac{10^9 \times 7}{10^9 + 7} \approx 7.00 \text{ days}$$

So, as one half-time gets very long relative to the other, the effective half-time approaches the *shorter of the two*.

Now, with the effective half-time concept included, we can conclude our analysis of the cumulative dose:

$$D = \int_0^T \dot{D}\,\mathrm{d}t = \frac{k \sum_i y_i E_i \phi_i}{m} \int_0^T A\,\mathrm{d}t$$

$$\int_0^T A\,\mathrm{d}t = \int_0^T f A_0 e^{-\lambda_e t}\,\mathrm{d}t = \frac{f A_0}{\lambda_e}(1 - e^{-\lambda_e T})$$

If we integrate the expression from 0 to ∞ instead of to a finite time T, the integral becomes just $f \times A_0/\lambda_e = f \times A_0 \times 1.443 \times T_e$.

$$D = \frac{k \int_0^\infty A\,\mathrm{d}t \sum_i y_i E_i \phi_i}{m}$$

$$D = \frac{k\tilde{A} \sum_i y_i E_i \phi_i}{m}$$

$$D = \frac{k\, 1.443 f A_0 T_e \sum_i y_i E_i \phi_i}{m}$$

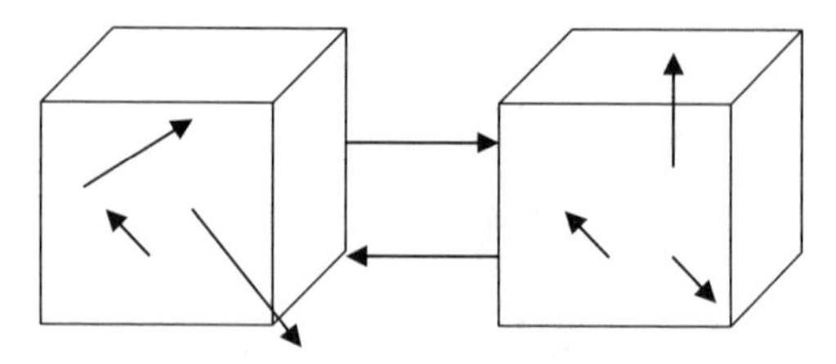

FIGURE 2.2. Two objects with a uniform distribution of radioactivity.

Now consider that we have two objects that contain a uniform distribution of radioactive material and are able to irradiate themselves, each other, and possibly other objects in the system (Fig. 2.2).

To get the total dose to either object (or any other object in the system, by extension), we need to add up all of the individual contributions. This is not more complicated conceptually; it just involves more terms:

$$D_1 = \frac{k\tilde{A}_1 \sum_i y_i E_i \phi_i (1 \leftarrow 1)}{m_1} + \frac{k\tilde{A}_2 \sum_i y_i E_i \phi_i (1 \leftarrow 2)}{m_1} + \ldots$$

$$D_2 = \frac{k\tilde{A}_1 \sum_i y_i E_i \phi_i (2 \leftarrow 1)}{m_2} + \frac{k\tilde{A}_2 \sum_i y_i E_i \phi_i (2 \leftarrow 2)}{m_2} + \ldots$$

Dosimetry Systems

The equations derived above are generic dose equations. Many authors and groups have developed this equation in one form or another to apply to different situations. Often, some of the factors in the equations are grouped together to simplify calculations, particularly when dealing with radionuclides with complex emission spectra. Some of the physical quantities (e.g., absorbed fraction and mass) may be combined into single values. However these quantities may be grouped, hidden, or otherwise moved around in different systems, all of them incorporate the concepts from these equations, and *all are based on the same basic concepts and principles.* Given the same input data and assumptions, one will obtain identical results. Sometimes, the apparent differences between the systems and their complicated-appearing equations may confuse and intimidate the user, who may be frustrated in trying to make any two of them agree for a given problem. Careful investigation to discern these grouped factors can help to resolve apparent differences. Let us try to understand each of the systems and see how they are equivalent.

Marinelli/Quimby Method

Publications by Marinelli et al. and Quimby and Feitelberg[2,3] gave the dose from a beta emitter that decays completely in a given organ or tissue as

$$D_\beta = 73.8\, C E_\beta T$$

where D_β is the beta dose in rad, C is the concentration of the nuclide in the tissue in microcuries per gram, E_β is the mean energy emitted per decay of the nuclide, and T is the half-life of the nuclide in the tissue. We know that the cumulated activity is given as 1.443 times the half-life times the initial activity in the tissue. The other terms in the equation are as follows: $k = (73.8/1.443) = 51.1$; C is activity per mass; and for beta emitters, as noted above, we assume that ϕ is 1.0. For gamma emitters, values of ϕ were estimated from the geometrical factors of Hine and Brownell[4] for spheres and cylinders of fixed sizes. Dose rates were based on expressions for dose near a point-source gamma emitter integrated over the source volume:

$$D_\gamma = 10^{-3}\, \Gamma C \int_v \frac{e^{-\mu r}}{r^2}\, dV \frac{\text{rad}}{\text{h}}$$

This equation does not seem to exactly fit the form of our general equation, but it does. The factor C is the activity per unit mass. The specific gamma rate constant Γ essentially gives the exposure rate per disintegration into an infinite medium from a point-source (equivalent to $k \times \Sigma\, n_i \times E_i$ in our generic equation). Finally, the factor $[\int \exp(-\mu r)/r^2 \, dV]$ is like an absorbed fraction: μ is an absorption coefficient and $1/r^2$ is essentially a geometric absorbed fraction. The integral in this expression can be obtained analytically only for simple geometries. Solutions for several standard objects (spheres, cylinders, and so forth) were provided in the geometric factors in Hine and Brownell's text.[4]

The International Commission on Radiological Protection

The ICRP has developed two comprehensive internal dosimetry systems intended for use in protecting radiation workers, mainly the nuclear fuel cycle, but certainly applicable as well to nuclear medicine workers. ICRP publication II[5] was the basis for the first set of complete radiation protection regulations in the United States (Code of Federal Regulations (CFR), Title 10, Chapter 20, or 10 CFR 20). These regulations were only replaced (completely) in 1994 when a revision of 10 CFR 20 incorporated many of the new procedures and results of the ICRP 30 series.[6] Even though written by the same scientific group, these two systems appear to be completely different. We have already noted, however, that they must be completely identical in concept and thus differ only in certain internal assumptions. Both of these systems, dealing with occupational exposures, were used to calculate dose equivalent instead of just absorbed dose.

In the ICRP II system, the dose equivalent rate was given by the expression:

$$H = \frac{51.2 A \xi}{m}$$

This looks somewhat like our original equation, converted to dose equivalent, but several factors appear to be missing. Recall we noted that terms are often grouped together for convenience, and that is true here. The missing factors are included in the term ξ:

$$\xi = \sum_i y_i E_i \phi_i Q_i$$

The factor 51.2 is k, which gives results in units of rem per day, for activity in microcuries, mass in grams, and energy in MeV. The ICRP developed a system of limitation of concentrations in air and water for employees from this equation and assumptions about the kinetic behavior of radionuclides in the body. These were the well-known maximum permissible concentrations (MPCs). Employees could be exposed to these concentrations on a continuous basis and not receive an annual dose rate to the so-called critical organ that exceeded established annual dose limits.

In the ICRP 30 system, the cumulative dose equivalent was given as:

$$H_{50,\mathrm{T}} = 1.6 \times 10^{-10} \sum_{\mathrm{S}} U_{\mathrm{S}}\,\mathrm{SEE}(\mathrm{T} \leftarrow \mathrm{S})$$

In this equation, T represents a target region and S represents a source region.

This equation looks altogether new; it seems that nothing much is similar to any of the equations we have looked at. This is simply, however, the same old equation wearing a new disguise. The factor SEE is merely:

$$\mathrm{SEE} = \frac{\sum_i y_i E_i \phi_i(\mathrm{T} \leftarrow \mathrm{S})\, Q_i}{m_{\mathrm{T}}}$$

The factor U_{S} is another symbol for cumulated activity, and the factor 1.6×10^{-10} is k. Note that the symbol Q (quality factor), used in some of the early ICRP manuals, is shown here instead of the current notation w_{R} (radiation weighting factor). In this system (based on the Système International unit system), this value of k produces cumulative dose equivalents in sievert, from activity in becquerels, mass in grams, energy in MeV, and appropriate quality factors. As in ICRP II, this equation was used to develop a system of dose limitation for workers, but, unlike the ICRP II system, limits are placed on *activity intake during a year, which would prevent cumulative doses* (not continuous dose rates) from exceeding established limits. These quantities of activity were called annual limits on intake (ALIs); derived air concentrations (DACs), which are directly analogous to MPCs for air, were calculated from ALIs. More recent ICRP documents (e.g., ICRP 71[7]) changed the formulation somewhat. For example, the equivalent dose at age t in target organ or tissue T due to an intake of a radionuclide at age t_0 may be given as:

$$H^{\cdot}(t, t_0) = \sum_{S} q_S(t, t_0)\,\mathrm{SEE}(\mathrm{T} \leftarrow \mathrm{S}; t)$$

where $q_S(t, t_0)$ is the activity of the radionuclide in source organ S at age t after intake at age t_0 (Bq), and $\mathrm{SEE}(\mathrm{T} \leftarrow \mathrm{S}, t)$ is the specific effective energy, as defined above, except that the energy is given in J and mass is given in kg, so the units of dose are Gy without the need for a unit conversion factor.

A real innovation in the ICRP 30 system is the so-called effective dose equivalent (H_e or EDE). Certain organs or organ systems were assigned dimensionless weighting factors that are a function of their assumed relative radiosensitivity for expressing fatal cancers or genetic defects. The assumed radiosensitivities were derived from the observed rates of expression of these effects in various populations exposed to radiation. Multiplying an organ's dose equivalent by its assigned weighting factor gives a weighted dose equivalent. The sum of weighted dose equivalents for a given exposure to radiation is the effective dose equivalent. Because of the way in which the weighting factors are derived, this effective dose equivalent is the dose equivalent that, if uniformly received by the whole body, would result in the same *risk* as from the individual organs receiving these different dose equivalents. It is entirely different from the dose equivalent to the whole body that is calculated using values of SEE for the total body. Whole-body doses are often meaningless in internal dose situations because nonuniform and localized energy deposition is averaged over the mass of the whole body (70 kg). More detail will be given in Chapter 3.

Another difference between doses calculated with the ICRP II system and the ICRP 30 (and Medical Internal Radiation Dose; MIRD) system is that the authors of ICRP II used a very simplistic phantom to estimate their absorbed fractions. All body organs and the whole body were represented as spheres of uniform composition. Furthermore, organs could only irradiate themselves, not other organs. So, although contributions from all emissions were considered, an organ could only receive a dose if it contained activity, and the absorbed fractions for photons were different from those calculated from the more advanced phantoms used by the ICRP 30 and MIRD systems.

Medical Internal Radiation Dose System

The equation for absorbed dose given in the Medical Internal Radiation Dose (MIRD) system[8] is

$$D_{r_k} = \sum_{h} \tilde{A}_h S(r_k \leftarrow r_h)$$

In this equation, r_k represents a target region, and r_h represents a source region. The use of the subscripts "h" and "k" for "source" and "target" is unusual. One might ask, why not the subscripts "s" and "t," as in the ICRP system? The reason for this is a bit amusing: the early work donc with this system was done by FORTRAN programmers. In the old FORTRAN, integers, which were used as looping indices, began in the alphabet with the letter "i." The letters "i" and "j" had already been used for other variables, so the letters "h" and "k" were used here (with "h" assigned to be an integer)! The cumulated activity is as defined above; all other terms were lumped in the factor "S":

$$S(r_k \leftarrow r_h) = \frac{k \sum_i y_i E_i \phi_i(r_k \leftarrow r_h)}{m_{r_k}}$$

In the MIRD equations, the factor k is 2.13, which gives doses in rad, from activity in microcuries, mass in grams, and energy in MeV. The MIRD system was developed primarily for use in estimating radiation doses received by patients from administered radiopharmaceuticals; it was not intended to be applied to a system of dose limitation for workers. To learn internal dosimetry for radiopharmaceuticals, Dr. Carol Marcus said that she "... took the Great Red Pamphlets home and began learning from them."[9] These "great red pamphlets" of the MIRD system indeed defined the methods, equations, and models for nuclear medicine dosimetry for many years. A listing of selected MIRD pamphlets is given in Table 2.2.

The MIRD Committee also produced other journal publications, books, and other resources that were vital to the day-to-day use of dosimetry in nuclear medicine. At present, however, most of the useful MIRD documents have fallen out of date and have been replaced with newer and more modern materials, including many software and Internet-based resources. The committee continues, however, to occasionally produce scientific documents of general interest to the internal dosimetry community.

RAdiation Dose Assessment Resource

In the early 21st century, an electronic resource was established on the Internet to provide rapid, worldwide dissemination of important dose quantities and data. The RAdiation Dose Assessment Resource (RADAR) established a Web site (www.doseinfo-radar.com) and provided a number of publications on the data and methods used in the system.

Table 2.2. Selected MIRD pamphlets.

Pamphlet	Publication date	Main information	Comments
1, 1 revised	1968, 1976	Discussion of MIRD internal dose technique	Superseded by the MIRD Primer (1988)
3	1968	Photon absorbed fractions for small objects	Superseded by J Nucl Med 41:149–160, 2000
5, 5 revised	1969, 1978	Description of anthropomorphic phantom representing Reference Man, photon absorbed fractions for many organs	Superseded by availability of Cristy/Eckerman phantom series (1987)
7	1971	Dose distribution around point sources, electron, beta emitters	Good data, difficult to use; use of Monte Carlo codes like Monte Carlo N-Particle (MCNP), Electron Gamma Shower (EGS) is generally preferred
8	1971	Photon absorbed fractions for small objects	Same as Pamphlet 3, smaller objects, also superseded by J Nucl Med 41:149–160, 2000
11	1975	*S* values for many nuclides	Newer *S* values available, see RADAR dose factor page
12	1977	Discussion of kinetic models for internal dosimetry	

(Continued)

TABLE 2.2. *(Continued)*

Pamphlet	Publication date	Main information	Comments
13	1981	Description of model of the heart, photon absorbed fractions	
14, 14 revised	1992, 1999	Dynamic urinary bladder for absorbed dose calculations	Software available, see RADAR software page
15	1996	Description of model for the brain, photon absorbed fractions	
16	1999	Outline of best practices and methods for collecting and analyzing kinetic data	Widely cited, useful document
17	1999	*S* values for voxel sources	
18	2001	Administered activity for xenon studies	
19	2003	Multipart kidney model with absorbed fractions	

The RADAR system[10] has perhaps the simplest representation of the cumulative dose equation:

$$D = N \times \mathrm{DF}$$

where N is the number of disintegrations that occur in a source organ, and DF is

$$\mathrm{DF} = \frac{k \sum_i y_i E_i \phi_i}{m}$$

The DF is conceptually similar to the "*S* value" defined in the MIRD system. The number of disintegrations is the integral of a time-activity curve for a source region. RADAR members produced compendia of decay data, dose conversion factors, and catalogued standardized dose models for radiation workers and nuclear medicine patients, among other resources. They also produced the widely used OLINDA/EXM[11] personal computer software code, which used the equations shown here and the input data from the RADAR site. This code was basically a revised version of the highly popular MIRDOSE[12] software, which implemented the MIRD method for internal dose calculations (but was not in any way associated with the MIRD Committee itself). The RADAR site and OLINDA/EXM software implement all of the most current and widely accepted models and methods for internal dose calculations (as are described in the next chapter) and are constantly updated to reflect changes that occur in the science of internal dose assessment.

RADAR is now an officially sanctioned committee, like MIRD and the ICRP, and its members have published a number of documents, data sets, and tools with a literature basis that is clearly important to the current practice of dosimetry. Some of the pertinent references are summarized in Table 2.3.

Usefulness of the Dosimetry Systems

Any of the systems above will give useful estimates of absorbed dose, assuming that good input data is provided to the system and assuming that appropriate models are employed for the dose conversion factors (see Chapter 3). The choice of a particular system is based on practicality and applicability. Most of the results in the ICRP systems are oriented toward protection of radiation workers, whereas those of the MIRD system are oriented towards nuclear medicine patients. The RADAR system is designed to accommodate either. Further, the RADAR system has been implemented in automated electronic methods that have

TABLE 2.3. Selected RADAR member articles.

Title	Authors	Publication information	Comments
Specific absorbed fractions of energy at various ages from internal photons sources	Cristy M and Eckerman K	ORNL/TM-8381 V1-V7, 1987	Absorbed fractions for a pediatric phantom series
Mathematical models and specific absorbed fractions of photon energy in the nonpregnant adult female and at the end of each trimester of pregnancy	Stabin M, Watson E, Cristy M, Ryman J, Eckerman K, Davis J, Marshall D, and Gehlen K	ORNL Report ORNL/TM 12907, 1995	Absorbed fractions for the pregnant female
MIRDOSE: personal computer software for internal dose assessment in nuclear medicine	Stabin M	J Nucl Med 37(3): 538–546, 1996	Description of the MIRDOSE software. Between the third and sixth most cited article in J Nucl Med history for many months
Radiation dose estimates for radiopharmaceuticals	Stabin MG, Stubbs JB, and Toohey RE	NUREG/CR-6345, 1996	Dose estimates for adults for a number of radiopharmaceuticals

Radiation absorbed dose to the embryo/fetus from radiopharmaceuticals	Russell JR, Stabin MG, Sparks RB, and Watson EE	Health Phys 73(5): 756–769, 1997	
Electron absorbed fractions and dose conversion factors for marrow and bone by skeletal regions	Eckerman K and Stabin M	Health Phys 78(2): 199–214, 2000	Bone/marrow dose model for children and adults
Re-evaluation of absorbed fractions for photons and electrons in small spheres	Stabin MG and Konijnenberg M	J Nucl Med 41: 149–160, 2000	New sphere absorbed fractions
New decay data for internal and external dose assessment	Stabin MG and da Luz CQPL	Health Phys 83(4): 471–475, 2002	Decay data for >800 radionuclides, used to develop RADAR dose factors
Physical models and dose factors for use in internal dose assessment	Stabin MG and Siegel JA	Health Phys 85(3): 294–310, 2003	Dose factors for >800 radionuclides and 15 phantoms

Abbreviations: ORNL, Oak Ridge National Laboratory.

been tested and used by the international nuclear medicine community for many years and are more complete and practical than those developed so far by the MIRD group (see Chapter 3). Many wish to use the old 1970s MIRD pamphlets still today, as they respect the authority of this group, sanctioned by the Society of Nuclear Medicine. Others regularly use the RADAR system, due to its convenience and wide acceptance.†

References

1. International Commission on Radiological Protection. 1990 Recommendations of the International Commission on Radiological Protection. ICRP Publication 60. Pergamon Press, New York, 1991.
2. Marinelli L, Quimby E, Hine G. Dosage determination with radioactive isotopes II, practical considerations in therapy and protection. Am J Roent Radium Ther 59:260–280, 1948.
3. Quimby E, Feitelberg S. Radioactive Isotopes in Medicine and Biology. Lea and Febiger, Philadelphia, 1963.
4. Hine G, Brownell G. Radiation Dosimetry. Academic Press, New York, 1956.
5. International Commission on Radiological Protection. Report of committee II on permissible dose for internal radiation. Health Phys 3, 1960.
6. International Commission on Radiological Protection. Limits for Intakes of Radionuclides by Workers. ICRP Publication 30. Pergamon Press, New York, 1979.
7. International Commission on Radiological Protection. Age-dependent Doses to Members of the Public from Intake of Radionuclides: Part 4, Inhalation Dose Coefficients. ICRP Publication 71. Pergamon Press, New York, 1996.
8. Loevinger R, Budinger T, Watson E. MIRD Primer for Absorbed Dose Calculations. Society of Nuclear Medicine, New York, 1988.
9. Stabin M. Nuclear medicine dosimetry. Phys Med Biol 51: R187–R202, 2006.
10. Stabin MG, Siegel JA. Physical models and dose factors for use in internal dose assessment. Health Phys 85:294–310, 2003.
11. Stabin MG, Sparks RB, Crowe E. OLINDA/EXM: The second-generation personal computer software for internal

† The article describing the MIRDOSE software was for many years between the third and sixth most cited journal article in the history of the *Journal of Nuclear Medicine*, as catalogued on the journal's Web site.

dose assessment in nuclear medicine. J Nucl Med 46: 1023–1027, 2005.
12. Stabin MG. MIRDOSE: personal computer software for internal dose assessment in nuclear medicine. J Nucl Med 37:538–546, 1996.

3
Models and Resources for Internal Dose Calculations

Standardized Models for Dosimetry

Reliable estimates of radiation dose from the use of diagnostic or therapeutic radiopharmaceuticals in nuclear medicine are essential to the evaluation of the risks and benefits of their use. To estimate absorbed dose for all significant tissues, one must determine for each tissue the quantity *absorbed dose*, which is the amount of energy from ionizing absorbed per unit mass of any material. Here our interest is in the dose to human tissues. One may also calculate *equivalent dose*, which includes terms describing the effectiveness of different radiations in producing biological damage in human tissues. It is important that investigators use *standardized* methods, models, and tools for the calculation of absorbed doses, so that other researchers, users, regulators, and others can readily understand and (if desired) reproduce the calculations.

We saw in the previous chapter that a generic equation for the absorbed dose rate in an object containing a uniform distribution of radioactivity (e.g., an organ or tissue with radiopharmaceutical uptake) may be shown as:

$$\dot{D}_{\mathrm{T}} = \frac{kA_{\mathrm{S}} \sum_i y_i E_i \phi_i(\mathrm{T} \leftarrow \mathrm{S})}{m_{\mathrm{T}}}$$

where $\dot{D}_T$ is absorbed dose rate to a target region of interest (Gy/s or rad/h), A_S is activity (MBq or μCi) in source region S, y_i is number of radiations with energy E_i emitted per nuclear transition, E_i is energy per radiation for the ith radiation (MeV), $\phi_i(T \leftarrow S)$ is fraction of energy emitted in a source region that is absorbed in a target region, m_T is mass of the target region (kg or g), and k is a proportionality constant (Gy-kg/MBq-s-MeV or rad-g/μCi-h-MeV).

The proportionality constant is a combination of all factors that are needed to obtain the dose rate in the desired units from the units employed for the other variables, and it is essential that this factor be properly calculated and applied. The results of our calculation will be useless unless the units within are consistent and they correctly express the quantity desired. Further details on the use and implementation of this equation will now be described.

For example, if we want the dose rate in Gy/s, and we have employed units of MBq for activity, MeV for energy, and kg for mass, the conversions that are needed are

$$k = \frac{10^6\,\text{dis}}{\text{s-MBq}}\,\frac{\text{Gy-kg}}{1\,\text{J}}\,\frac{1.6\times10^{-13}\,\text{J}}{1\,\text{MeV}} = 1.6\times10^{-7}\,\frac{\text{Gy-kg}}{\text{MBq-s-MeV}}$$

If we want the dose rate in rad/h, and we have employed units of μCi for activity, MeV for energy, and g for mass, the conversions that are needed are

$$k = \frac{3.7\times10^4\,\text{dis}}{\text{s-}\mu\text{Ci}}\,\frac{3600\,\text{s}}{\text{h}}\,\frac{\text{rad-g}}{100\,\text{erg}}\,\frac{1.6\times10^{-6}\,\text{erg}}{1\,\text{MeV}} = 2.13\,\frac{\text{rad-g}}{\mu\text{Ci-h-MeV}}$$

The application of *radiation weighting factors* (formerly called *quality factors*) to this equation to calculate the *equivalent dose* rate may also be included:

$$\dot{H} = \frac{kA_S \sum_i y_i E_i \phi_i w_{R_i}}{m_T}$$

Here, $\dot{H}$ is equivalent dose rate to a target region of interest (Sv/s or rem/h), and w_{R_i} is the radiation weighting factor assigned to the ith radiation.

Often, investigators usually are interested in an estimate of *total absorbed dose* rather than just the instantaneous dose rate at some point in time from a radiopharmaceutical administration. In this equation, the quantity activity (nuclear transitions per unit time) causes the outcome of the equation to have a time dependence. To calculate cumulative dose, the time integral of the dose equation must be calculated. In most cases, the only term that depends on time is activity, so the only factor that has to be integrated is the activity term. There are exceptions to this rule: tumors may shrink during therapy, and changes in the size of the thyroid gland during hyperthyroidism therapy may affect the calculation of dose, [1] for example. The integral of the time-activity curve (i.e., the area under that curve, regardless of its shape) is often called the *cumulated activity* (often given with the symbol $\tilde{A}$), and it represents the total *number of disintegrations that have occurred over time in a source region.*

Therefore, the equation for cumulative dose is

$$D_{\mathrm{T}} = \int \dot{D}_{\mathrm{T}} \mathrm{d}t = \frac{k\tilde{A}_{\mathrm{S}} \sum_{i} y_i E_i \phi_i}{m_{\mathrm{T}}}$$

where D is the absorbed dose (Gy or rad), and the quantity $\tilde{A}_{\mathrm{S}}$ represents the integral of $A_{\mathrm{S}}(t)$, the time-dependent activity within the source region:

$$\tilde{A}_{\mathrm{S}} = \int_0^{\infty} A_{\mathrm{S}}(t)\,\mathrm{d}t = A_0 \int_0^{\infty} f_{\mathrm{S}}(t)\,\mathrm{d}t$$

where A_0 is the activity administered to the patient at time $t = 0$, and $f_{\mathrm{S}}(t)$ is the *fractional distribution function* for source region r_{S} (fraction of administered activity present within the source region r_{S} at time t). The quantity $\tilde{A}_{\mathrm{S}}$ can thus be considered the *activity-time integral,* or ATI. In many instances, the function $f_{\mathrm{S}}(t)$ may be modeled as a sum of exponential functions:

$$f_{\mathrm{S}}(t) = f_1 e^{-(\lambda_1+\lambda_{\mathrm{p}})t} + f_2 e^{-(\lambda_2+\lambda_{\mathrm{p}})t} + \cdots + f_N e^{-(\lambda_N+\lambda_{\mathrm{p}})t}$$

where terms $f_1 \ldots f_N$ represent the fractional uptake of the administered activity within the in source region r_S, $\lambda_1 \ldots \lambda_N$ represent the biological elimination constants for these same compartments, and λ_P represents the physical decay constant for the radionuclide of interest. Uptake phases in an organ will be represented by negative values of f_i, whereas elimination phases will be represented by positive values of f_i. The source region retention is most often characterized by exponential functions, but other functions as well may describe the uptake and removal of activity.

In normalizing by the administered activity, the MIRD system wrote in terms of the *fractional residence time* τ_S in source region r_S:

$$\tau_S = \frac{\tilde{A}_S}{A_0} = \int_0^\infty \frac{A_S(t)}{A_0} \, dt = \int_0^\infty \alpha_S(t) \, dt$$

The mean absorbed dose to target region r_T per unit administered activity A_0 has been given as[2]:

$$\hat{D}(r_T) = \frac{\bar{D}(r_T)}{A_0} = \sum_S \tau_S S(r_T \leftarrow r_S)$$

where $\tilde{A}$ is defined as above, τ was defined in the MIRD system as the "residence time," which represented the ratio $\tilde{A}/A_0$, the cumulated activity divided by the activity administered to the subject (A_0), and S is given by:

$$S = \frac{k \sum_i y_i E_i \phi_i}{m_T}$$

This concept of "residence time,"[2] however, has often caused confusion because of its apparent units of time (even though it really expresses the number of nuclear transitions that occur in a source region (e.g., Bq-s) normalized to the activity administered (e.g., Bq) and because of the use of this term to represent the "mean life" of atoms in biological or engineering applications. Its use *should be replaced* by the *normalized cumulated activity* ($\tilde{A}/A_0$), which has units of (for example) Bq-s/Bq administered or μCi-h/μCi administered. As noted above, *cumulated activity* ($\tilde{A}$) represents the total number of disintegrations that have occurred in a source region over a given time of integration (Bq-s or μCi-h); therefore, the *normalized cumulated activity* ($\tilde{A}/A_0$) represents the total number of disintegrations that have occurred in a source region over a given time of integration per unit activity initially administered to a subject (Bq or μCi). The units of the quantity should not be thought of as time, and the confusion of it representing a mean time that an atom spends in a region (which it is not) is removed.

A generalized expression for calculating internal dose, which may describe the equations shown in publications by different authors, is given in the RADAR dose calculation system[3] by the following equation:

$$D = N \times \mathrm{DF}$$

where N is the number of nuclear transitions that occur in source region S, and DF is a "dose factor." The factor DF contains the various components shown in the formulas for S and SEE; basically, it depends on combining decay data with absorbed fractions (AFs), which are derived generally using Monte Carlo simulation of radiation transport in models of the body and its internal structures (organs, tumors, etc.):

$$\mathrm{DF} = \frac{k \sum_i y_i E_i \phi_i w_{R_i}}{m_{\mathrm{T}}}$$

As written, the equations above give only the dose from one source region to one target region, but they can be generalized easily to multiple source and target regions.

$$D_{\mathrm{T}} = \frac{k \sum_S \tilde{A}_{\mathrm{S}} \sum_i y_i E_i \phi_i (\mathrm{T} \leftarrow \mathrm{S})}{m_{\mathrm{T}}}$$

The RADAR calculational system was implemented in the OLINDA/EXM software code,[4] as mentioned in Chapter 2 and to be described in more detail later in this chapter.

The Effective Dose Concept

The ICRP, in its 1979 description of radiation protection quantities and limits for radiation workers,[5] defined a new dosimetry quantity, the *effective dose equivalent* (H_e, or EDE). The ICRP subsequently renamed this quantity *effective dose* (E) in 1991.[6] Certain organs or organ systems were assigned dimensionless weighting factors (Table 3.1), which are a function of their assumed relative radiosensitivity for expressing fatal cancers or genetic defects.

The assumed radiosensitivities were derived from the observed rates of expression of these effects in various populations exposed to radiation. Multiplying an organ's dose equivalent by its assigned weighting factor gives a weighted dose equivalent. The *sum of weighted dose equivalents* for a given exposure to radiation is the effective dose:

$$E = \sum_{T} H_{\mathrm{T}} \times w_{\mathrm{T}}$$

Here is an example (using ICRP 30 weighting factors):

Organ	Actual dose equivalent received (Sv)		Weighting factor		Weighted dose equivalent (Sv)
Gonads	0.0010	×	0.25	=	0.00025
Breast	0.0020	×	0.15	=	0.00030
Lungs	0.0020	×	0.12	=	0.00024
Red marrow	0.0015	×	0.12	=	0.00018
Thyroid	0.0005	×	0.03	=	0.000015
Bone surfaces	0.0020	×	0.03	=	0.00006
Liver	0.0030	×	0.06	=	0.00018
	Sum = Effective dose equivalent			=	0.0012 Sv

The effective dose is meant to represent the equivalent dose, which, if received uniformly by the whole body, would result in the same total risk as that actually incurred by a given actual nonuniform irradiation. It is *entirely different* from the dose equivalent to the "whole body" that is calculated using values of SEE for the total body. "Whole-body"

TABLE 3.1. Weighting factors recommended by the ICRP for calculation of the effective dose.

Organ	ICRP 30	ICRP 60	New[a]
Gonads	0.25	0.20	0.08
Red marrow	0.12	0.12	0.12
Colon		0.12	0.12
Lungs	0.12	0.12	0.12
Stomach		0.12	0.12
Bladder		0.05	0.05
Breasts	0.15	0.05	0.08
Liver		0.05	0.05
Esophagus		0.05	0.05
Thyroid	0.03	0.05	0.05
Skin		0.01	0.01
Bone surfaces	0.03	0.01	0.01
Salivary glands, brain			0.01
Remainder	0.30	0.05	0.12

Sources: Data in first column from International Commission on Radiological Protection. Limits for Intakes of Radionuclides by Workers. ICRP Publication 30. Pergamon Press, New York, 1979. Data in second column from International Commission on Radiological Protection. 1990 Recommendations of the International Commission on Radiological Protection. ICRP Publication 60. Pergamon Press, New York, 1991.

[a] Proposed, for possible release in 2007.

doses are basically meaningless in nuclear medicine applications, as nonuniform and localized energy deposition is simply averaged over the mass of the whole body (70 kg). Thus, if a radiopharmaceutical concentrates heavily in a few organs, all of the energy absorbed by these (and other) organs is divided by the mass of the whole body to obtain the "whole-body" dose, which is not meaningful. Table 3.2 summarizes some of the dose quantities of interest in nuclear medicine dosimetry.

Some have objected to the use of the effective dose quantity in nuclear medicine, because of the uncertainties involved and the fact that the quantity was derived for use with a radiation worker population.[7] The ICRP itself, however, as well as many other international organizations, has affirmed that the quantity is useful for nuclear medicine

TABLE 3.2. Summary of nuclear medicine dose quantities.

Quantity	Units	Comments
Individual organ dose (absorbed dose or equivalent dose)	Gy or Sv	Doses to all available organs and tissues in the standardized phantoms should be routinely reported.
Maximum dose organ (absorbed dose or equivalent dose)	Gy or Sv	The individual organ that receives the highest dose per unit activity administered or per study should be considered in study design and execution.
Whole-body dose (absorbed dose or equivalent dose)	Gy or Sv	Useful *only* if all organs and tissues in the body receive an approximately uniform dose. Rarely of value for radiopharmaceuticals. Most useful in external dose assessment.
Effective dose	Sv	Risk weighted effective whole-body dose. Gives the equivalent dose uniform to the whole body that theoretically has the same risk as the actual, nonuniform dose pattern received.

applications, the associated uncertainties notwithstanding. It is clearly more useful in evaluating and comparing doses between radiopharmaceuticals with different distribution and retention patterns in the body. It is *very important*, however, to use and interpret this quantity correctly:

- The quantity should *never be used in situations involving radiation therapy,* as it is related to the evaluation of stochastic risks from exposures involving low doses and dose rates.
- It should *not be used to evaluate the risk to a given individual*; its application is to populations that receive doses at these levels.

If one accepts the quantity, with all of its inherent assumptions and uncertainties, however, it provides some useful features:

- As just noted, it allows direct comparison of different radiopharmaceuticals that may have completely different radiation dose patterns. For example, compare the use of ^{201}Tl chloride with ^{99m}Tc Sestamibi for use in myocardial imaging studies. There are many variables that enter into a discussion of which agent is preferable for these studies, and we will not review all of them here. But, just from a radiation dose standpoint, if one uses, for example, 74 MBq (2 mCi) of ^{201}Tl chloride, the two highest dose organs are the thyroid, which may receive about 40 mGy (4 rad), and the kidneys, which may receive about 30 mSv (3 rem).[8] One might instead use 740 MBq (20 mCi) of ^{99m}Tc Sestamibi, in which case the two highest dose organs are the gallbladder, which may receive about 29 mSv (2.9 rem), and the kidneys, which may receive about 27 mSv (2.7 rem) (rest patients).[9] The kidney doses are similar, but is 40 mGy to the thyroid more acceptable than 29 mGy to the gallbladder? The effective doses for ^{201}Tl chloride is 11.5 mSv (1.15 rem) and for ^{99m}Tc Sestamibi it is 6.7 mSv (0.67 rem). So, strictly from a dose standpoint, the use of ^{99m}Tc Sestamibi appears more desirable, although this was not immediately obvious by looking at the highest dose organs.
- Effective doses from radiopharmaceuticals may be added to those received from other procedures outside of nuclear medicine. For example, if a typical value of an effective dose for a lumbar spine X-ray is 0.7 mSv (0.07 rem), and a subject has had two such exams recently and then receives a ^{99m}Tc Sestamibi heart scan, the total effective dose is estimated as $6.7 + (2 \times 0.7) = 8.1$ mSv (0.81 rem).
- A popular way to explain radiation risks in a simple way that many members of the public can understand is to express the dose in terms of equivalent years of exposure to background radiation.[10] Estimates of background radiation dose rates vary, but if one chooses 3 mSv/year (300 mrem/year) as an example, then the ^{99m}Tc

Sestamibi study discussed above may be thought of as equivalent in total risk to slightly more than 2 years of exposure to natural background radiation.

Input Data for Dose Calculations

The input data needed for a numerical solution of the equations above include three major components:

1. Decay data for the radionuclide in question (values of y_i and E_i).
2. Biokinetic data for the radiopharmaceutical under study (α_i and T_e values).
3. Absorbed fractions (ϕ) for different radiations and all source and target regions of interest.
4. Masses of the target regions (m_T).

A number of sources of radionuclide *decay data* are available in paper or electronic form: for example, the 1989 MIRD compendium,[11] ICRP 38,[12] and the 2002 RADAR compendium[13] (Table 3.3). Excellent online sources are also available:

- JAERI (http://www.jaeri.go.jp/)
- Brookhaven National Laboratory (http://www.nndc.bnl.gov/)
- National Institute of Standards and Technology (http://physics. nist.gov/PhysRefData/contents.html)
- Lund University (http://nucleardata.nuclear.lu.se/nuclear data/toi/index.asp)

TABLE 3.3. Features of some decay data compendia.

Compendium	No. nuclides	Decay scheme diagrams?	Alpha emitters?	Online?
MIRD 1989	> 240	Yes	No	No
ICRP 38	> 800	Yes	Yes	No
RADAR	> 800	No	Yes	Yes

The RADAR compendium was compiled from data on the BNL site.

Biokinetic data are established via observations, in animal or human subjects, of the distribution and clearance of radiopharmaceuticals from different regions of the body. The last *two parameters* (ϕ and m_T) are obtained from reference models of the human body that have been developed and then used with radiation transport simulations to calculate absorbed fraction values.

Anthropomorphic Phantoms

The current generation of anthropomorphic phantoms began with the development of the Fisher-Snyder phantom.[14] This phantom used a combination of geometric shapes—spheres, cylinders, cones, and so forth—to create a more anatomically accurate representation of the body. Monte Carlo computer programs were used to simulate the creation and transport of photons through these various structures in the body whose atomic composition and density were based on data provided in the ICRP report on "Reference Man."[15] This report provided various anatomic data assumed to represent the average working adult male in the Western hemisphere and has been recently updated.[16] Although this was most often applied to adult males, this phantom also contained regions representing organs specific to the adult female. Using this phantom, radiation doses were calculated for adults based on activity residing in any organ and irradiating any other organ. Absorbed fractions at discrete photon energies were calculated and published.[14] In addition, S values, as defined above, for more than 100 radionuclides and for more than 20 source and target regions were also published.[17]

Cristy-Eckerman Child and Adult Phantoms

Cristy and Eckerman[18] modified the adult male model somewhat, but, more importantly, developed models for a series of individuals of different size and age. Six phantoms were developed, which were assumed to represent children of ages 0 (newborn), 1 year, 5 years, 10 years, and 15 years, and adults of both sexes. Absorbed fractions for photons at discrete energies were published for these phantoms, which contained approximately 25 source and target regions. Tables of *S* values were never published, but ultimately they were made available in the MIRDOSE computer software.[19]

Models of the Pregnant Female

Stabin et al. developed a series of phantoms for the adult female, both nonpregnant and at three stages of pregnancy.[20] These phantoms modeled the changes to the uterus, intestines, bladder, and other organs that occur during pregnancy and included specific models for the fetus, fetal soft tissue, fetal skeleton, and placenta. *S* values for these phantoms were also made available through the MIRDOSE software.[19]

Bone and Marrow Models

Spiers et al. at the University of Leeds[21] first established electron absorbed fractions (AFs) for bone and marrow in a healthy adult male, which were used in the dose factors (DFs), or *S* values, in MIRD Pamphlet No. 11.[17] Eckerman reevaluated this work and extended the results to derive DFs for 15 skeletal regions in six models representing individuals of various ages.[22] The results were used in the MIRDOSE 3 software[19] to provide mean marrow dose, regional marrow dose, and dose-volume histograms for different individuals. Bouchet et al.[23] used updated information on regional bone and marrow mass and calculated new AFs using the EGS4 Monte Carlo code. Although the results of the latter two efforts are similar in most characteristics and reported results, the models differed in three areas:

1. Eckerman multiplied AFs in the marrow space (MS) by the fractional cellularity (cellularity factor; CF) of a given bone to obtain red marrow AFs, but Bouchet et al. did not.
2. Eckerman assumed that surface-distributed sources of bone-seeking nuclides reside on an infinitely thin layer adjacent to the bone mineral, whereas Bouchet et al. assumed that they were distributed by volume within the 10-μm layer of tissue representing the dividing cells on the surfaces of growing bone.
3. Electrons entering the 10-μm layer of tissue on bone surfaces were assumed by Eckerman to have entry angles distributed randomly chosen between 0 and 180 degrees, while Bouchet et al. chose angles from a cosine distribution.

These differences resulted in significant differences in the reported AFs and DFs in some cases. Recently, a revised model has been derived[24] that resolves these model differences in ways best supported by currently available data. In this model, marrow-to-marrow electron absorbed fractions were derived using the Bouchet et al. values at low energies and the Eckerman values at medium-to-high energies. The rationale for this is that, at very low energies, when the electron range is comparable with the dimensions of the cells in the marrow, the electron absorbed fractions should approach 1.0, and not the CF in that bone, as in the Bouchet et al. model. As the electron energy increases and the electrons traverse many cell diameters or even marrow cavities before depositing all of their energy, the fraction of total energy deposited in the active marrow should be the CF times the fraction deposited in the total MS (as in the Eckerman model) and not simply the absorbed fraction in the MS (as in the Bouchet et al. model). The skeletal average active marrow to active marrow AFs for the revised model are shown in Figure 2.1 in the previous chapter. As discussed in the article by Stabin et al.,[24] new data were only available[25] to guide the modifications for the adult model. Nonetheless, adjustments were made as well to the values of AF(RM ← RM) for the pediatric models, assuming that the values approached 1.0 at low energies and followed similar trends with energy as were noted in the adult. A revision has also been implemented to the cortical bone AFs reported by Eckerman and Stabin.[22] Skeletal average AFs for all bone regions employed in the calculations in this study were given. Values of AF(RM ← RM) and values of AF(BS ← CB-S) and AF(BS ← CB-V) differ from those originally published by Eckerman and Stabin,[22] for the reasons discussed earlier.

Currently Available Dose Conversion Factors

For many years, the only source of dose factors for use in practical calculations were found in MIRD Pamphlet No. 11,[17] in which factors were given for about 25 organs, but only in the adult male phantom, for 117 radionuclides. The MIRDOSE code[19] provided dose factors for more than 240 radionuclides, for about 25 organs as well, but in the entire Cristy-Eckerman and Stabin et al. pediatric, adult, and pregnant female phantoms series (10 phantoms). Stabin and Siegel[3] calculated dose factors for more than 800 radionuclides for:

1. All source and target regions in the six models in the Cristy-Eckerman phantom series.[18]
2. All source and target regions in the four models in the Stabin et al. pregnant female phantoms series.[20]
3. All target regions in the Watson et al. peritoneal cavity model.[26]
4. All target regions in the Stabin prostate gland model.[27]
5. All source and target regions in the six models of the MIRD head and brain model.[28]
6. All source and target regions in the MIRD regional kidney model.[29]
7. The unit density sphere models of Stabin and Konijnenberg.[30]

These dose factors are based on decay data from the Brookhaven National Laboratory resource (http://www.nndc.bnl.gov/) and are useful for implementation in the dose equations described earlier. These dose conversion factors use the child, adult, and pregnant woman phantoms and bone and marrow models described above and included standard modeling assumptions, as were described in that paper.

Organ Masses

The masses of the target regions in the full-body phantoms described above are shown in Tables 3.4 and 3.5. These values have been published previously but are repeated here, as they are important to the calculation of the DFs and are often needed for making modifications to the DFs or understanding the numerical values. A sample table of DFs for ^{99m}Tc is shown in Table 3.6. New standard organ masses have been recently recommended by the ICRP[16] and are currently being used to design new phantoms for use in dose calculations. These masses, and all DFs, will be updated in future publications, as the data become available.

Standard Dose Estimates for Radiopharmaceuticals

The approaches and models described above may be used by many individuals and groups to calculate dose estimates for different radiopharmaceuticals. It can sometimes be frustrating to some users to seek dosimetry information on a particular radiopharmaceutical and find several different sets of dose estimates, often with minor and sometimes significant discrepancies in the models employed and the resulting doses calculated. Standardized dose calculations for a handful of radiopharmaceuticals (less than 20) were developed by the MIRD committee over a 30-year period. Others were published at one time by the dosimetry information center at Oak Ridge.[31] Most recently, a large compendium of dose estimates has been published by a working group of the ICRP for around 200 radiopharmaceuticals, based on the best known biodistribution data and using the standard

TABLE 3.4. Total body dimensions and masses of source regions in the Cristy-Eckerman phantom series.

Phantom:	Newborn	1 year	5 years	10 years	15 years	Adult
Phantom height (cm):	47.5	69.6	103	132	157	167
Body surface area (cm^2)[a]:	2100	3900	7500	9600	13,300	18,000
Organ			Mass (g) of organ			
Adrenals	5.83	3.52	5.27	7.22	10.5	16.3
Brain	352	884	1260	1360	1410	1420
Breasts including skin	0.205	1.1	2.17	3.65	407	403
Breasts excluding skin	0.107	0.732	1.51	2.6	361	351
Gallbladder contents	2.12	4.81	19.7	38.5	49	55.7
Gallbladder wall	0.408	0.91	3.73	7.28	9.27	10.5
GI tract						
LLI contents	6.98	18.3	36.6	61.7	109	143
LLI wall	7.96	20.6	41.4	70	127	167
Small intestine contents	52.9	138	275	465	838	1100

Stomach contents	10.6	36.2	75.1	133	195	260
Stomach wall	6.41	21.8	49.1	85.1	118	158
ULI contents	11.2	28.7	57.9	97.5	176	232
ULI wall	10.5	27.8	55.2	93.4	168	220
Heart contents	36.5	72.7	134	219	347	454
Heart wall	25.4	50.6	92.8	151	241	316
Kidneys	22.9	62.9	116	173	248	299
Liver	121	292	584	887	1400	1910
Lungs	50.6	143	290	453	651	1000
Ovaries	0.328	0.714	1.73	3.13	10.5	8.71
Pancreas	2.8	10.3	23.6	30	64.9	94.3
Remaining tissue[b]	2360	6400	13,300	23,100	40,000	51,800
Skeleton						
Active marrow	47	150	320	610	1050	1120
Cortical bone	0	299	875	1580	3220	4000
Trabecular bone	140	200	219	396	806	1000
Skin	118	271	538	888	2150	3010
Spleen	9.11	25.5	48.3	77.4	123	183
Testes	0.843	1.21	1.63	1.89	15.5	39.1
Thymus	11.3	22.9	29.6	31.4	28.4	20.9
Thyroid	1.29	1.78	3.45	7.93	12.4	20.7

(Continued)

TABLE 3.4. *(Continued)*

Phantom:	Newborn	1 year	5 years	10 years	15 years	Adult
Phantom height (cm):	47.5	69.6	103	132	157	167
Body surface area (cm^2)[a]:	2100	3900	7500	9600	13,300	18,000
Organ			Mass (g) of organ			
Urinary bladder contents	12.4	32.9	64.7	103	160	211
Urinary bladder wall	2.88	7.7	14.5	23.2	35.9	47.6
Uterus	3.85	1.45	2.7	4.16	79	79
Whole body	3600	9720	19,800	33,200	56,800	73,700

Source: Adapted with permission from Cristy M, Eckerman K. Specific absorbed fractions of energy at various ages from internal photons sources. ORNL/TM-8381 V1-V7. Oak Ridge National Laboratory, Oak Ridge, TN, 1987. With permission of the Oak Ridge National Laboratory, managed by UT-Battelle, LLC, for the U.S. Department of Energy.

[a] As suggested in International Commission on Radiological Protection. Report of the Task Group on Reference Man. ICRP Publication 23. Pergamon Press, New York, 1975.

[b] "Remaining tissue" is defined as the part of the phantom remaining when all defined organs have been removed. This region of the phantom has been used in the radiation transport code to model muscle for dosimetric purposes. The appropriate masses of muscle to use in such calculations are newborn 760 g; 1-year-old 1000 g; 5-year-old 2000 g; 10-year-old 7000 g; 15-year-old 15,500; adult male 28,000 g.[15]

Abbreviations: LLI, lower large intestine; SI, small intestine; ULI, upper large intestine.

TABLE 3.5. Masses of source regions in the pregnant female phantom series.

Phantom:	Adult female (nonpregnant)	Three-month pregnant female	Six-month pregnant female	Nine-month pregnant female
Organ	Mass (g) of organ			
Adrenals	14	14	14	14
Brain	1200	1200	1200	1200
Breasts excluding skin	360	360	360	360
Gallbladder contents	50	50	50	50
Gallbladder wall	8	8	8	8
GI tract				
LLI contents	135	135	135	135
LLI wall	160	160	160	160
Small intestine contents	375	375	375	375
Small intestine wall	600	600	600	600
Stomach contents	230	230	230	230

(Continued)

TABLE 3.5. *(Continued)*

Phantom:	Adult female (nonpregnant)	Three-month pregnant female	Six-month pregnant female	Nine-month pregnant female
Organ	Mass (g) of organ			
Stomach wall	140	140	140	140
ULI contents	210	210	210	210
ULI wall	200	200	200	200
Heart contents	410	410	410	410
Heart	240	240	240	240
Kidneys	275	275	275	275
Liver	1400	1400	1400	1400
Lungs	800[b]	800[b]	800[b]	800[b]
Ovaries	11	11	11	11
Pancreas	85	85	85	85
Remaining tissue[a]	40,000	39,300	41,700	39,500
Skeleton				
Active marrow	1050[c]	1050[c]	1050[c]	1050[c]
Cortical bone	3000	3000	3000	3000
Trabecular bone	750	750	750	750
Skin	1790	1790	1790	1790
Spleen	150	150	150	150

Thymus	20	20	20	20
Thyroid	17	17	17	17
Urinary bladder contents	160	128	107	42.3
Urinary bladder wall	35.9	36.9	34.5	23.9
Uterine wall	80	374	834	1095
Fetus	—	458	1640	2960
Placenta	—	—	310	466
Whole body	58,000	58,000	61,500	63,700
Whole body (maternal tissues)	56,800	56,400	57,500	56,600

Source: Adapted with permission from Stabin M, Watson E, Cristy M, Ryman J, Eckerman K, Davis J, Marshall D, Gehlen K. Mathematical models and specific absorbed fractions of photon energy in the nonpregnant adult female and at the end of each trimester of pregnancy. ORNL Report ORNL/TM-12907. Oak Ridge National Laboratory, Oak Ridge, TN, 1995. With permission of the Oak Ridge National Laboratory, managed by UT-Battelle, LLC, for the U.S. Department of Energy.

[a] "Remaining tissue" is defined as the part of the phantom remaining when all defined organs have been removed. This region of the phantom has been used in the radiation transport code to model muscle for dosimetric purposes. However, the appropriate mass of muscle to use in such calculations in the adult female is 17,000 g.

[b] Although the mass reported in the phantom[20] is 651 g, the value of 800 g recommended in the ICRP "Reference Man" report[15] is used for dose calculations.

[c] Although the mass reported in the phantom[20] is 1300 g, the value of 1050 g recommended in the Cristy and Eckerman model[18] for the 15-year-old is used for dose calculations

Abbreviations: LLI, lower large intestine; ULI, upper large intestine.

TABLE 3.6. Dose factors (mGy/MBq-s) for all organs in the adult male phantom for ^{99m}Tc.

	Source organ						
Target organ	Adrenals	Brain	Breasts	GB Cont	LLI Cont	SI Cont	StomCont
Adrenals	1.80E-04	4.18E-10	5.05E-08	3.13E-07	2.25E-08	7.46E-08	2.73E-07
Brain	4.18E-10	4.23E-06	3.17E-09	1.49E-10	1.57E-11	3.91E-11	4.27E-10
Breasts	5.05E-08	3.17E-09	1.14E-05	3.33E-08	2.28E-09	7.35E-09	5.73E-08
Gallbladder wall	3.57E-07	1.54E-10	3.41E-08	3.37E-05	6.49E-08	4.38E-07	3.05E-07
LLI wall	1.98E-08	1.32E-11	2.42E-09	5.93E-08	1.23E-05	5.92E-07	9.10E-08
Small intestine	7.46E-08	3.91E-11	7.35E-09	4.58E-07	7.16E-07	4.22E-06	2.08E-07
Stomach wall	2.85E-07	2.52E-10	5.93E-08	2.93E-07	1.24E-07	2.13E-07	8.53E-06
ULI wall	9.41E-08	4.76E-11	7.51E-09	7.78E-07	3.10E-07	1.36E-06	2.65E-07
Heart wall	2.85E-07	2.54E-09	2.61E-07	1.04E-07	5.42E-09	2.06E-08	2.33E-07
Kidneys	7.24E-07	1.58E-10	1.99E-08	3.89E-07	7.10E-08	2.13E-07	2.73E-07
Liver	4.35E-07	8.16E-10	6.82E-08	8.20E-07	1.80E-08	1.16E-07	1.47E-07
Lungs	2.33E-07	7.63E-09	2.33E-07	7.09E-08	4.50E-09	1.35E-08	1.10E-07
Muscle	1.12E-07	2.21E-08	4.25E-08	1.14E-07	1.23E-07	1.12E-07	9.96E-08
Ovaries	3.14E-08	1.52E-11	2.61E-09	1.11E-07	1.26E-06	9.23E-07	5.85E-08
Pancreas	1.09E-06	4.15E-10	6.22E-08	6.75E-07	5.21E-08	1.42E-07	1.23E-06
Red marrow	2.53E-07	1.01E-07	5.52E-08	1.02E-07	2.01E-07	1.79E-07	7.50E-08
Osteogenic cells	2.67E-07	2.99E-07	7.76E-08	1.14E-07	1.82E-07	1.49E-07	1.03E-07
Skin	3.41E-08	3.97E-08	7.63E-08	3.09E-08	3.62E-08	3.01E-08	3.41E-08
Spleen	4.58E-07	5.19E-10	4.37E-08	1.33E-07	6.53E-08	1.01E-07	7.83E-07
Testes	1.54E-09	1.46E-12	0.00E+00	6.91E-09	1.40E-07	2.61E-08	2.90E-09

Thymus	5.66E-08	6.88E-09	2.29E-07	1.43E-08	2.04E-09	4.66E-09	3.65E-08
Thyroid	8.11E-09	1.35E-07	3.01E-08	2.45E-09	2.48E-10	4.87E-10	2.62E-09
Urinary bladder wall	7.55E-09	6.02E-12	1.33E-09	4.22E-08	4.98E-07	2.12E-07	1.73E-08
Uterus	1.89E-08	1.31E-11	2.62E-09	1.16E-07	5.17E-07	8.37E-07	5.05E-08
Total body	1.72E-07	1.25E-07	1.03E-07	1.36E-07	1.49E-07	1.59E-07	1.17E-07

Target organ	ULI Cont	HeartCon	Hrt Wall	Kidneys	Liver	Lungs	Muscle
Adrenals	9.58E-08	2.53E-07	2.85E-07	7.24E-07	4.35E-07	2.33E-07	1.12E-07
Brain	4.68E-11	3.14E-09	2.54E-09	1.58E-10	8.16E-10	7.63E-09	2.21E-08
Breasts	8.00E-09	2.41E-07	2.61E-07	1.99E-08	6.82E-08	2.33E-07	4.25E-08
Gallbladder wall	7.53E-07	1.03E-07	1.22E-07	4.09E-07	8.70E-07	7.46E-08	1.19E-07
LLI wall	2.14E-07	4.06E-09	4.90E-09	5.50E-08	1.44E-08	3.29E-09	1.34E-07
Small intestine	1.25E-06	1.57E-08	2.06E-08	2.13E-07	1.16E-07	1.35E-08	1.12E-07
Stomach wall	2.86E-07	1.66E-07	2.65E-07	2.53E-07	1.48E-07	1.19E-07	1.09E-07
ULI wall	8.37E-06	2.12E-08	2.65E-08	2.12E-07	1.88E-07	1.81E-08	1.10E-07
Heart wall	2.97E-08	5.48E-06	1.19E-05	8.22E-08	2.33E-07	4.40E-07	9.20E-08
Kidneys	2.12E-07	6.45E-08	8.22E-08	1.32E-05	2.93E-07	6.66E-08	9.79E-08
Liver	1.87E-07	2.13E-07	2.33E-07	2.93E-07	3.16E-06	1.97E-07	7.52E-08
Lungs	1.77E-08	4.59E-07	4.40E-07	6.66E-08	2.09E-07	3.57E-06	9.36E-08
Muscle	1.07E-07	8.83E-08	9.20E-08	9.79E-08	7.52E-08	9.34E-08	1.93E-07
Ovaries	7.71E-07	4.55E-09	6.15E-09	7.02E-08	3.81E-08	5.39E-09	1.44E-07
Pancreas	1.62E-07	2.65E-07	3.57E-07	4.97E-07	3.86E-07	1.74E-07	1.24E-07
Red marrow	1.43E-07	1.11E-07	1.11E-07	1.71E-07	8.32E-08	1.11E-07	9.07E-08

(Continued)

Table 3.6. *(Continued)*

Target organ	ULI Cont	HeartCont	Hrt Wall	Kidneys	Liver	Lungs	Muscle
Osteogenic cells	1.27E-07	1.60E-07	1.60E-07	1.62E-07	1.24E-07	1.66E-07	1.84E-07
Skin	3.09E-08	3.41E-08	3.70E-08	3.79E-08	3.62E-08	4.02E-08	5.72E-08
Spleen	1.05E-07	1.24E-07	1.67E-07	6.63E-07	7.22E-08	1.64E-07	1.03E-07
Testes	1.92E-08	5.16E-10	6.16E-10	3.10E-09	1.57E-09	3.67E-10	9.89E-08
Thymus	5.43E-09	8.87E-07	7.35E-07	1.73E-08	5.93E-08	2.85E-07	1.06E-07
Thyroid	7.69E-10	5.17E-08	4.33E-08	2.95E-09	8.64E-09	8.82E-08	1.16E-07
Urinary bladder wall	1.61E-07	2.22E-09	2.17E-09	1.87E-08	1.16E-08	1.33E-09	1.40E-07
Uterus	3.97E-07	4.87E-09	5.47E-09	6.42E-08	3.29E-08	4.10E-09	1.43E-07
Total body	1.41E-07	1.17E-07	1.65E-07	1.58E-07	1.59E-07	1.44E-07	1.33E-07

Target organ	Ovaries	Pancreas	Red Mar.	CortBoneS	TrabBoneS	CortBoneV	TrabBoneV
Adrenals	3.14E-08	1.09E-06	2.41E-07	1.07E-07	1.07E-07	1.07E-07	1.07E-07
Brain	1.52E-11	4.15E-10	8.08E-08	1.17E-07	1.17E-07	1.17E-07	1.17E-07
Breasts	2.61E-09	6.22E-08	5.12E-08	3.10E-08	3.10E-08	3.10E-08	3.10E-08
Gallbladder wall	9.91E-08	8.14E-07	1.15E-07	4.41E-08	4.41E-08	4.41E-08	4.41E-08
LLI wall	1.12E-06	4.17E-08	1.98E-07	7.25E-08	7.25E-08	7.25E-08	7.25E-08
Small intestine	9.23E-07	1.42E-07	1.86E-07	5.74E-08	5.74E-08	5.74E-08	5.74E-08
Stomach wall	5.85E-08	1.26E-06	8.23E-08	3.89E-08	3.89E-08	3.89E-08	3.89E-08
ULI wall	8.29E-07	1.69E-07	1.55E-07	5.02E-08	5.02E-08	5.02E-08	5.02E-08
Heart wall	6.15E-09	3.57E-07	1.09E-07	5.74E-08	5.74E-08	5.74E-08	5.74E-08
Kidneys	7.02E-08	4.97E-07	1.70E-07	6.22E-08	6.22E-08	6.22E-08	6.22E-08
Liver	3.81E-08	3.86E-07	8.96E-08	4.82E-08	4.82E-08	4.82E-08	4.82E-08

Lungs	5.39E-09	1.76E-07	1.09E-07	6.67E-08	6.67E-08	6.67E-08	6.67E-08
Muscle	1.44E-07	1.24E-07	9.07E-08	7.45E-08	7.45E-08	7.45E-08	7.45E-08
Ovaries	3.23E-04	3.65E-08	2.13E-07	6.54E-08	6.54E-08	6.54E-08	6.54E-08
Pancreas	3.65E-08	3.74E-05	1.48E-07	6.54E-08	6.54E-08	6.54E-08	6.54E-08
Red marrow	2.13E-07	1.40E-07	1.79E-06	2.01E-07	6.38E-07	2.01E-07	3.88E-07
Osteogenic cells	1.66E-07	1.67E-07	1.01E-06	2.73E-06	3.15E-06	7.26E-07	1.21E-06
Skin	3.09E-08	3.01E-08	4.20E-08	4.68E-08	4.68E-08	4.68E-08	4.68E-08
Spleen	3.85E-08	1.28E-06	9.19E-08	4.94E-08	4.94E-08	4.94E-08	4.94E-08
Testes	0.00E+00	2.58E-09	3.09E-08	4.09E-08	4.09E-08	4.09E-08	4.09E-08
Thymus	1.94E-09	6.13E-08	8.45E-08	4.83E-08	4.83E-08	4.83E-08	4.83E-08
Thyroid	2.29E-10	7.28E-09	7.54E-08	7.67E-08	7.67E-08	7.67E-08	7.67E-08
Urinary bladder wall	5.49E-07	1.38E-08	9.18E-08	4.01E-08	4.01E-08	4.01E-08	4.01E-08
Uterus	1.57E-06	3.73E-08	1.54E-07	4.89E-08	4.89E-08	4.89E-08	4.89E-08
Total body	1.87E-07	1.85E-07	1.52E-07	1.40E-07	1.40E-07	1.40E-07	1.40E-07

	Source organ						
Target organ	Spleen	Testes	Thymus	Thyroid	UB Cont	Uterus	TotBody
Adrenals	4.58E-07	1.54E-09	5.66E-08	8.11E-09	8.41E-09	1.89E-08	1.67E-07
Brain	5.19E-10	1.46E-12	6.88E-09	1.35E-07	5.94E-12	1.31E-11	1.21E-07
Breasts	4.37E-08	0.00E+00	2.29E-07	3.01E-08	1.30E-09	2.62E-09	1.00E-07
Gallbladder wall	1.35E-07	6.71E-09	2.65E-08	2.64E-09	3.45E-08	1.15E-07	1.77E-07
LLI wall	4.65E-08	1.96E-07	1.61E-09	2.06E-10	5.78E-07	4.97E-07	1.73E-07
Small intestine	1.01E-07	2.61E-08	4.66E-09	4.87E-10	2.24E-07	8.37E-07	1.77E-07
Stomach wall	7.75E-07	4.43E-09	3.61E-08	3.71E-09	2.10E-08	5.53E-08	1.58E-07

(Continued)

TABLE 3.6. *(Continued)*

Target organ	Source organ						
	Adrenals	Brain	Breasts	GB Cont	LLI Cont	SI Cont	StomCont
ULI wall	1.06E-07	1.78E-08	5.31E-09	7.69E-10	1.60E-07	4.17E-07	1.71E-07
Heart wall	1.67E-07	6.16E-10	7.35E-07	4.33E-08	2.22E-09	5.47E-09	1.61E-07
Kidneys	6.63E-07	3.10E-09	1.73E-08	2.95E-09	2.00E-08	6.42E-08	1.54E-07
Liver	7.22E-08	1.57E-09	5.93E-08	8.64E-09	1.17E-08	3.29E-08	1.55E-07
Lungs	1.65E-07	3.67E-10	2.97E-07	8.82E-08	1.04E-09	4.10E-09	1.41E-07
Muscle	1.03E-07	9.89E-08	1.06E-07	1.16E-07	1.30E-07	1.43E-07	1.31E-07
Ovaries	3.85E-08	0.00E+00	1.94E-09	2.29E-10	5.41E-07	1.57E-06	1.81E-07
Pancreas	1.28E-06	2.58E-09	6.13E-08	7.28E-09	1.38E-08	3.73E-08	1.79E-07
Red marrow	8.43E-08	2.69E-08	8.25E-08	7.94E-08	8.02E-08	1.39E-07	1.34E-07
Osteogenic cells	1.26E-07	1.02E-07	1.23E-07	1.97E-07	1.05E-07	1.30E-07	3.89E-07
Skin	3.49E-08	1.03E-07	4.39E-08	4.38E-08	3.90E-08	3.01E-08	9.13E-08
Spleen	2.25E-05	2.17E-09	3.89E-08	7.83E-09	8.40E-09	2.57E-08	1.54E-07
Testes	2.17E-09	8.66E-05	1.91E-10	2.31E-11	3.73E-07	0.00E+00	1.28E-07
Thymus	3.89E-08	1.91E-10	1.51E-04	1.62E-07	8.06E-10	1.81E-09	1.43E-07
Thyroid	7.83E-09	2.31E-11	1.62E-07	1.50E-04	9.56E-11	2.14E-10	1.45E-07
Urinary bladder wall	8.04E-09	3.85E-07	8.14E-10	9.65E-11	1.10E-05	1.28E-06	1.67E-07
Uterus	2.57E-08	0.00E+00	1.81E-09	2.14E-10	1.24E-06	4.70E-05	1.83E-07
Total body	1.59E-07	1.32E-07	1.48E-07	1.49E-07	1.18E-07	1.88E-07	1.39E-07

Note: Produced using OLINDA/EXM code and selecting technetium-99m.

Abbreviations: GB Cont, gallbladder contents; LLI Cont, lower large intestine contents; ULI Cont, upper large intestine contents; SI Cont, small intestine contents; StomCont, stomach contents; HeartCon, heart contents; HrtWall, heart wall; Red Mar., red marrow; CortBoneS, cortical bone surfaces; TrabBoneS, trabecular bone surfaces; CortBoneV, cortical bone volume; TrabBoneV, trabecular bone volume; UB Cont, urinary bladder contents; TotBody, total body.

models described earlier.[32] These values should be referenced in most cases in which standardized dose estimates for a particular radiopharmaceutical are needed.

Modifications to the Standard Models

Individual Variations in Organ Mass

Some efforts are under way to provide true patient-specific dose calculations for individual patients in therapy, based on anatomic and physiologic data taken from each patient. Methods for such approaches are not yet standardized and are still a matter for research. Fairly simple modifications can be made, however, to the standard equations shown earlier in cases in which the mass of an individual's organ is known to be significantly different than that of the standardized phantom used. For alpha and beta emissions, a linear scaling of dose with mass is appropriate, as the absorbed fraction for emissions when the source is the target is just 1.0, and thus the DF just changes inversely with changes in mass of the organ. That is

$$\mathrm{DF}_2 = \mathrm{DF}_1 \frac{m_1}{m_2}$$

Here, DF_1 and DF_2 are the dose factors appropriate for use with organ masses m_1 and m_2.

For photons, Snyder[33] showed that the photon absorbed fractions vary directly with the cube root of the mass for self-irradiation (i.e., source = target), if the photon mean path length is large compared with the organ diameter, and that they vary directly with the mass for cross-irradiation (i.e., source ≠ target). What the latter point shows is that the *specific* absorbed fraction for cross-irradiation does not change with differences in mass, provided the source and target are sufficiently separated and that the change in mass of one or both does not appreciably change the distance between them. Thus, for self-irradiation, the absorbed fraction increases with the cube root of the mass of the organ, and thus the *specific* absorbed fraction decreases with the 2/3 power of the mass[33]:

$$\phi_2 = \phi_1 \left(\frac{m_2}{m_1}\right)^{1/3} \qquad \Phi_2 = \Phi_1 \left(\frac{m_1}{m_2}\right)^{2/3}$$

This relationship is useful but not necessarily exactly true for all body regions and radionuclides.[34]

Source in Hollow Organ Wall

The activity in hollow organs (e.g., GI tract organs, gallbladder, urinary bladder) is almost always assumed to be in the organ contents, with the target being cells in the organ wall. Although there is interest in calculating doses as a function of depth into the wall, the usual approach is to average the electron and photon doses over the mass of the wall. The electron dose is usually estimated as the maximum dose at the contents/wall interface, which is known to be calculated with a specific absorbed fraction of

$$\Phi(\text{wall} \leftarrow \text{contents})_{\text{electrons}} = \frac{1}{2 \times m_{\text{contents}}}$$

where m_{contents} is the mass of that organ's contents. Occasionally, the situation arises in which the activity in one of these organs is known to be concentrated in the wall of the organ, not in the contents. The specific absorbed fractions for the photons will not vary much whether the source is in the wall or in the contents, so we assume that these do not need adjustment. The dose to the organ in a real case arises from a number of contributions, from the organ to itself and to the organ from other sources. These contributions need to be separated from the electron contribution from the organ's contents to the walls, corrected, then added back together to perform this correction appropriately. An example will be given in the Teaching Examples in Chapters 4 and 5.

The Pregnant Patient

An area of particular concern in nuclear medicine is the pregnant or potentially pregnant patient. A 1997 document in the journal *Health Physics*[35] gave estimates of fetal dose from more than 80 radiopharmaceuticals, based on some standardized kinetic models and, in some cases, including knowledge of the amount of radiopharmaceutical crossover, as measured in animal or human studies. Summary tables of dose estimates for many important radiopharmaceuticals are shown in Table 3.7. Note that the estimates for ^{18}F FDG have been updated since 1997, as new information was found on the placental crossover of this compound.[36]

Some special cases are sometimes encountered as well for the pregnant patient:

1. Fetal thyroid dose: If radioiodine is administered to a woman who has passed about 10 to 13 weeks of gestation, the fetal thyroid will have been formed, and this tiny organ concentrates the iodine that crosses the placenta. Evelyn Watson calculated doses to the fetal thyroid per unit activity administered to the mother.[37] Her results are presented in Table 3.8. (The doses are in mGy to the fetal thyroid per MBq administered to the mother.)
2. The hyperthyroid patient: Fetal dose has not been well established for patients whose iodine kinetics differ from the standard model for iodine-131 NaI. In early pregnancy (when most of these exposures should occur, as the therapy will be clearly contraindicated in patients known to be pregnant), values from a 1991 *Journal of Nuclear Medicine* article[38] should serve well. Their estimates are given in Table 3.9.
3. The athyroid patient: In thyroid cancer patients, iodine-131 NaI is often given to patients whose thyroids have been mostly removed surgically. There may be a remnant of thyroid tissue and/or some thyroid cancer metastases around the body, but usually a large amount of activity is given (enough to destroy all remaining thyroid tissue and the mets). In a study involving a few athyroidic subjects,[39]

TABLE 3.7. Absorbed dose estimates to the embryo/fetus per unit activity of radiopharmaceutical administered to the mother (shading indicates maternal and fetal self dose contributions).

Radiopharmaceutical	Early mGy/MBq[a]	3 Month mGy/MBq	6 Month mGy/MBq	9 Month mGy/MBq
^{57}Co vitamin B-1, normal-flushing	1.0×10^{0}	6.8×10^{-1}	8.4×10^{-1}	8.8×10^{-1}
^{57}Co vitamin B-12, normal-no flushing	1.5×10^{0}	1.0×10^{0}	1.2×10^{0}	1.3×10^{0}
^{57}Co vitamin B-12, PA-flushing	2.1×10^{-1}	1.7×10^{-1}	1.7×10^{-1}	1.5×10^{-1}
^{57}Co vitamin B-12, PA-no flushing	2.8×10^{-1}	2.1×10^{-1}	2.2×10^{-1}	2.0×10^{-1}
^{58}Co vitamin B-12, normal-flushing	2.5×10^{0}	1.9×10^{0}	2.1×10^{0}	2.1×10^{0}
^{58}Co vitamin B-12, normal-no flushing	3.7×10^{0}	2.8×10^{0}	3.1×10^{0}	3.1×10^{0}
^{58}Co vitamin B-12, PA-flushing	8.3×10^{-1}	7.4×10^{-1}	6.4×10^{-1}	4.8×10^{-1}
^{58}Co vitamin B-12, PA-no flushing	9.8×10^{-1}	8.5×10^{-1}	7.6×10^{-1}	6.0×10^{-1}
^{60}Co vitamin B-12, normal-flushing	3.7×10^{1}	2.8×10^{1}	3.1×10^{1}	3.2×10^{1}
^{60}Co vitamin B-12, normal-no flushing	5.5×10^{1}	4.2×10^{1}	4.7×10^{1}	4.7×10^{1}
^{60}Co vitamin B-12, PA-flushing	5.9×10^{0}	4.7×10^{0}	4.8×10^{0}	4.5×10^{0}
^{60}Co vitamin B-12, PA-no flushing	8.3×10^{0}	6.5×10^{0}	6.8×10^{0}	6.5×10^{0}
^{18}F FDG[b]	2.2×10^{-2}	2.2×10^{-2}	1.7×10^{-2}	1.7×10^{-2}
^{18}F sodium fluoride	2.2×10^{-2}	1.7×10^{-2}	7.5×10^{-3}	6.8×10^{-3}
^{67}Ga citrate	9.3×10^{-2}	2.0×10^{-1}	1.8×10^{-1}	1.3×10^{-1}
^{123}I hippuran	3.1×10^{-2}	2.4×10^{-2}	8.4×10^{-3}	7.9×10^{-3}
^{123}I IMP	1.9×10^{-2}	1.1×10^{-2}	7.1×10^{-3}	5.9×10^{-3}
^{123}I MIBG	1.8×10^{-2}	1.2×10^{-2}	6.8×10^{-3}	6.2×10^{-3}

^{123}I sodium iodide	2.0×10^{-2}	1.4×10^{-2}	1.1×10^{-2}	9.8×10^{-3}
^{124}I sodium iodide	1.4×10^{-1}	1.0×10^{-1}	5.9×10^{-2}	4.6×10^{-2}
^{125}I HSA	2.5×10^{-1}	7.8×10^{-2}	3.8×10^{-2}	2.6×10^{-2}
^{125}I IMP	3.2×10^{-2}	1.3×10^{-2}	4.8×10^{-3}	3.6×10^{-3}
^{125}I MIBG	2.6×10^{-2}	1.1×10^{-2}	4.1×10^{-3}	3.4×10^{-3}
^{125}I sodium iodide	1.8×10^{-2}	9.5×10^{-3}	3.5×10^{-3}	2.3×10^{-3}
^{126}I sodium iodide	7.8×10^{-2}	5.1×10^{-2}	3.2×10^{-2}	2.6×10^{-2}
^{130}I sodium iodide	1.8×10^{-1}	1.3×10^{-1}	7.6×10^{-2}	5.7×10^{-2}
^{131}I hippuran	6.4×10^{-2}	5.0×10^{-2}	1.9×10^{-2}	1.8×10^{-2}
^{131}I HSA	5.2×10^{-1}	1.8×10^{-1}	1.6×10^{-1}	1.3×10^{-1}
^{131}I MAA	6.7×10^{-2}	4.2×10^{-2}	4.0×10^{-2}	4.2×10^{-2}
^{131}I MIBG	1.1×10^{-1}	5.4×10^{-2}	3.8×10^{-2}	3.5×10^{-2}
^{131}I sodium iodide	7.2×10^{-2}	6.8×10^{-2}	2.3×10^{-1}	2.7×10^{-1}
^{131}I rose bengal	2.2×10^{-1}	2.2×10^{-1}	1.6×10^{-1}	9.0×10^{-2}
^{111}In DTPA	6.5×10^{-2}	4.8×10^{-2}	2.0×10^{-2}	1.8×10^{-2}
^{111}In pentetreotide	8.2×10^{-2}	6.0×10^{-2}	3.5×10^{-2}	3.1×10^{-2}
^{111}In platelets	1.7×10^{-1}	1.1×10^{-1}	9.9×10^{-2}	8.9×10^{-2}
^{111}In red blood cells	2.2×10^{-1}	1.3×10^{-1}	1.1×10^{-1}	8.6×10^{-2}
^{111}In white blood cells	1.3×10^{-1}	9.6×10^{-2}	9.6×10^{-2}	9.4×10^{-2}
^{99m}Tc albumin microspheres	4.1×10^{-3}	3.0×10^{-3}	2.5×10^{-3}	2.1×10^{-3}
^{99m}Tc disofenin	1.7×10^{-2}	1.5×10^{-2}	1.2×10^{-2}	6.7×10^{-3}
^{99m}Tc DMSA	5.1×10^{-3}	4.7×10^{-3}	4.0×10^{-3}	3.4×10^{-3}
^{99m}Tc DTPA	1.2×10^{-2}	8.7×10^{-3}	4.1×10^{-3}	4.7×10^{-3}

(Continued)

TABLE 3.7. *(Continued)*

Radiopharmaceutical	Early mGy/MBq[a]	3 Month mGy/MBq	6 Month mGy/MBq	9 Month mGy/MBq
^{99m}Tc DTPA aerosol	5.8×10^{-3}	4.3×10^{-3}	2.3×10^{-3}	3.0×10^{-3}
^{99m}Tc glucoheptonate	1.2×10^{-2}	1.1×10^{-2}	5.3×10^{-3}	4.6×10^{-3}
^{99m}Tc HDP	5.2×10^{-3}	5.4×10^{-3}	3.0×10^{-3}	2.5×10^{-3}
^{99m}Tc HEDP	7.2×10^{-3}	5.2×10^{-3}	2.7×10^{-3}	2.4×10^{-3}
^{99m}Tc HMPAO	8.7×10^{-3}	6.7×10^{-3}	4.8×10^{-3}	3.6×10^{-3}
^{99m}Tc human serum albumin	5.1×10^{-3}	3.0×10^{-3}	2.6×10^{-3}	2.2×10^{-3}
^{99m}Tc MAA	2.8×10^{-3}	4.0×10^{-3}	5.0×10^{-3}	4.0×10^{-3}
^{99m}Tc MAG3	1.8×10^{-2}	1.4×10^{-2}	5.5×10^{-3}	5.2×10^{-3}
^{99m}Tc MDP	6.1×10^{-3}	5.4×10^{-3}	2.7×10^{-3}	2.4×10^{-3}
^{99m}Tc MIBI-rest	1.5×10^{-2}	1.2×10^{-2}	8.4×10^{-3}	5.4×10^{-3}
^{99m}Tc MIBI-stress	1.2×10^{-2}	9.5×10^{-3}	6.9×10^{-3}	4.4×10^{-3}
^{99m}Tc pertechnetate	1.1×10^{-2}	2.2×10^{-2}	1.4×10^{-2}	9.3×10^{-3}
^{99m}Tc PYP	6.0×10^{-3}	6.6×10^{-3}	3.6×10^{-3}	2.9×10^{-3}
^{99m}Tc RBC-heat treated	1.7×10^{-3}	1.6×10^{-3}	2.1×10^{-3}	2.2×10^{-3}
^{99m}Tc RBC-in vitro	6.8×10^{-3}	4.7×10^{-3}	3.4×10^{-3}	2.8×10^{-3}
^{99m}Tc RBC-in vivo	6.4×10^{-3}	4.3×10^{-3}	3.3×10^{-3}	2.7×10^{-3}
^{99m}Tc sulfur colloid-normal	1.8×10^{-3}	2.1×10^{-3}	3.2×10^{-3}	3.7×10^{-3}
^{99m}Tc sulfur colloid-liver disease	3.2×10^{-3}	2.5×10^{-3}	2.8×10^{-3}	2.8×10^{-3}

^{99m}Tc teboroxime	8.9×10^{-3}	7.1×10^{-3}	5.8×10^{-3}	3.7×10^{-3}
^{99m}Tc white blood cells	3.8×10^{-3}	2.8×10^{-3}	2.9×10^{-3}	2.8×10^{-3}
^{201}Tl chloride	9.7×10^{-2}	5.8×10^{-2}	4.7×10^{-2}	2.7×10^{-2}
^{127}Xe, 5-minute rebreathing, 5 liter spirometer volume	4.3×10^{-4}	2.4×10^{-4}	1.9×10^{-4}	1.5×10^{-4}
^{127}Xe, 5-minute rebreathing, 7.5 liter spirometer volume	2.3×10^{-4}	1.3×10^{-4}	1.0×10^{-4}	8.4×10^{-5}
^{127}Xe, 5 minute rebreathing, 10 liter spirometer volume	2.3×10^{-4}	1.4×10^{-4}	1.1×10^{-4}	9.2×10^{-5}
^{133}Xe, 5 minute rebreathing, 5 liter spirometer volume	4.1×10^{-4}	4.8×10^{-5}	3.5×10^{-5}	2.6×10^{-5}
^{133}Xe, 5-minute rebreathing, 7.5 liter spirometer volume	2.2×10^{-4}	2.6×10^{-5}	1.9×10^{-5}	1.5×10^{-5}
^{133}Xe, 5-minute rebreathing, 10 liter spirometer volume	2.5×10^{-4}	2.9×10^{-5}	2.1×10^{-5}	1.6×10^{-5}
^{133}Xe, injection	4.9×10^{-6}	1.0×10^{-6}	1.4×10^{-6}	1.6×10^{-6}

Source: Adapted with permission from Russell JR, Stabin MG, Sparks RB, Watson EE. Radiation absorbed dose to the embryo/fetus from radiopharmaceuticals. Health Phys 73:756–769, 1997.

[a] mGy/MBq $\times$ 3.7 $\equiv$ rad/mCi.

[b] Stabin M. Proposed addendum to previously published fetal dose estimate tables for 18F-FDG. J Nucl Med 45:634–635, 2004.

TABLE 3.8. Dose to the fetal thyroid (doses are mGy to the fetal thyroid per MBq administered to the mother).

Gestational age (months)	^{123}I	^{124}I	^{125}I	^{131}I
3	2.7	24	290	230
4	2.6	27	240	260
5	6.4	76	280	580
6	6.4	100	210	550
7	4.1	96	160	390
8	4.0	110	150	350
9	2.9	99	120	270

Source: Adapted with permission of ORAU from Watson EE. Radiation absorbed dose to the human fetal thyroid. In: Fifth International Radiopharmaceutical Dosimetry Symposium. Watson EE, Schlafke-Stelson, eds, Oak Ridge Associated Universities, Oak Ridge, TN, 1992, pp. 179–187.

it was found that the kinetics could be well characterized by treating the iodine not taken up by the thyroid by the normal kinetics of urinary bladder excretion (6.1-hour half-time). Using these assumptions and assuming that the other normal soft tissue uptakes occur, and using Russell's[35] results for fetal residence times (here it seems reasonable to assume that the standard kinetic model for maternal-fetal exchange of iodine would be similar to the euthyroid case), the dose estimates in Table 3.10 are

TABLE 3.9. Dose to the fetus in early pregnancy for hyperthyroid subjects (doses are in mGy to the fetus per MBq administered to the mother).

Maximum thyroid uptake	20%	40%	60%	80%	100%
"Fast" thyroid uptake[a]	0.049	0.044	0.040	0.036	0.036
"Normal" thyroid uptake[a]	0.063	0.058	0.055	0.052	0.053

Source: Adapted by permission of the Society of Nuclear Medicine from Stabin MG, Watson EE, Marcus CS, Salk RD. Radiation dosimetry for the adult female and fetus from iodine-131 administration in hyperthyroidism. J Nucl Med 32:808–813, 1991.

[a] "Fast" thyroid uptake meant an uptake half-time of 2.9 hours; "normal" meant a half-time of 6.1 hours.

obtained. Again, the dose estimates in later pregnancy are not likely to be of interest very often, as this kind of therapy should not be carried out on a pregnant woman.

4. Postadministration conception: An unusual kinetic picture sometimes arises when conception occurs after the iodine has been administered. In this case, the iodine has already started to wash out of the body, and whatever iodine is left will irradiate the embryo. This problem was evaluated,[40] and the results are shown in Tables 3.11 and 3.12.

The Breast-Feeding Patient

If a patient who is breast-feeding is administered a radiopharmaceutical, we are interested in how long we should interrupt breast-feeding (if at all) to protect the nursing infant. This subject has been studied by several authors. Most recently, a review was published in the *Journal of Nuclear Medicine*.[41] Table 3.13 shows the main recommendations of these authors.

TABLE 3.10. Dose to the fetus for athyroid patients (doses are in mGy to the fetus per MBq administered to the mother).

Early pregnancy	0.068
3 months	0.070
6 months	0.225
9 months	0.27

Source: Data from Rodriguez M. Development of a kinetic model and calculation of radiation dose estimates for sodium-iodide-^{131}I in athyroid individuals. Master's project, Colorado State University, 1996.

TABLE 3.11. Absorbed dose to fetus when conception occurs after administration, in cases of hyperthyroidism (mGy/MBq).

% Max uptake	Time in weeks after administration that conception occurs							
	1	2	3	4	5	6	7	8
5	4.1E-04	1.9E-04	8.7E-05	4.0E-05	1.9E-05	8.7E-06	4.0E-06	1.9E-06
10	8.3E-04	3.8E-04	1.7E-04	8.0E-05	3.7E-05	1.7E-05	7.8E-06	3.6E-06
15	1.3E-03	5.8E-04	2.6E-04	1.2E-04	5.5E-05	2.5E-05	1.1E-05	5.2E-06
20	1.7E-03	7.8E-04	3.5E-04	1.6E-04	7.2E-05	3.3E-05	1.5E-05	6.7E-06
25	2.2E-03	9.8E-04	4.4E-04	2.0E-04	8.8E-05	4.0E-05	1.8E-05	8.0E-06
30	2.7E-03	1.2E-03	5.3E-04	2.3E-04	1.0E-04	4.6E-05	2.0E-05	9.1E-06
35	3.2E-03	1.4E-03	6.1E-04	2.7E-04	1.2E-04	5.2E-05	2.3E-05	1.0E-05
40	3.7E-03	1.6E-03	7.0E-04	3.0E-04	1.3E-04	5.7E-05	2.4E-05	1.1E-05
45	4.3E-03	1.8E-03	7.8E-04	3.3E-04	1.4E-04	6.0E-05	2.6E-05	1.1E-05
50	4.8E-03	2.0E-03	8.5E-04	3.6E-04	1.5E-04	6.3E-05	2.6E-05	1.1E-05
55	5.4E-03	2.2E-03	9.2E-04	3.8E-04	1.6E-04	6.4E-05	2.7E-05	1.1E-05

60	6.0E-03	2.4E-03	9.8E-04	4.0E-04	1.6E-04	6.4E-05	2.6E-05	1.0E-05
65	6.7E-03	2.6E-03	1.0E-03	4.0E-04	1.6E-04	6.2E-05	2.5E-05	9.7E-06
70	7.3E-03	2.8E-03	1.1E-03	4.1E-04	1.5E-04	5.9E-05	2.2E-05	8.6E-06
75	7.9E-03	2.9E-03	1.1E-03	4.0E-04	1.5E-04	5.4E-05	2.0E-05	7.2E-06
80	8.5E-03	3.0E-03	1.1E-03	3.7E-04	1.3E-04	4.6E-05	1.6E-05	5.7E-06
85	9.1E-03	3.0E-03	1.0E-03	3.4E-04	1.1E-04	3.8E-05	1.3E-05	4.2E-06
90	9.6E-03	3.0E-03	9.2E-04	2.9E-04	8.9E-05	2.8E-05	8.6E-06	2.7E-06
95	9.8E-03	2.8E-03	7.9E-04	2.2E-04	6.3E-05	1.8E-05	5.1E-06	1.4E-06
100	9.8E-03	2.4E-03	6.1E-04	1.5E-04	3.8E-05	9.3E-06	2.3E-06	5.8E-07

Source: Adapted with permission of ORAU from Sparks RB, Stabin M. Fetal radiation dose estimates for I-131 sodium iodide in cases where accidental conception occurs after administration. In: Stelson A, Stabin M, Sparks R, eds. Sixth International Radiopharmaceutical Dosimetry Symposium. Oak Ridge Associated Unversities, Oak Ridge, TN, 1999, pp. 360–364.

TABLE 3.12. Absorbed dose to fetus when conception occurs after administration, in euthyroid cases (mGy/MBq).

% Max uptake	Time in weeks after administration that conception occurs							
	1	2	3	4	5	6	7	8
5	3.1E-04	1.5E-04	7.7E-05	3.8E-05	1.9E-05	9.5E-06	4.7E-06	2.4E-06
15	8.8E-04	4.4E-04	2.2E-04	1.1E-04	5.6E-05	2.8E-05	1.4E-05	7.2E-06
25	1.4E-03	7.1E-04	3.6E-04	1.8E-04	9.2E-05	4.7E-05	2.4E-05	1.2E-05

Source: Adapted with permission of ORAU from Sparks RB, Stabin M. Fetal radiation dose estimates for I-131 sodium iodide in cases where accidental conception occurs after administration. In: Stelson A, Stabin M, Sparks R, eds. Sixth International Radiopharmaceutical Dosimetry Symposium. Oak Ridge Associated Unversities, Oak Ridge, TN, 1999, pp. 360–364.

TABLE 3.13. Summary of recommendations regarding radiopharmaceuticals excreted in breast milk.

Radiopharmaceutical	Administered activity MBq (mCi)	Counseling[a]	Advice
^{67}Ga-citrate	185 (5)	Yes	Cessation
^{99m}Tc-DTPA	740 (20)	No	
^{99m}Tc-MAA	148 (4)	Yes	12 h
^{99m}Tc-pertechnetate	185 (5)	Yes	4 h
^{131}I-NaI	5550 (150)	Yes	Cessation
^{51}Cr-EDTA	1.85 (0.05)	No	
^{99m}Tc-DISIDA	300 (8)	No	
^{99m}Tc-glucoheptonate	740 (20)	No	
^{99m}Tc-HAM	300 (8)	No	
^{99m}Tc-MIBI	1110 (30)	No	
^{99m}Tc-MDP	740 (20)	No	
^{99m}Tc-PYP	740 (20)	No	
^{99m}Tc-RBC in vivo	740 (20)	Yes	12 h
^{99m}Tc-RBCs in vitro	740 (20)	No	
^{99m}Tc-sulfur colloid	444 (12)	No	
^{111}In-WBCs	18.5 (0.5)	No	
^{123}I-NaI	14.8 (0.4)	Yes	Cessation[b]
^{123}I-OIH	74 (2)	No	
^{123}I-MIBG	370 (10)	Yes	48 h
^{125}I-OIH	0.37 (0.01)	No	
^{131}I-OIH	11.1 (0.3)	No	
^{99m}Tc-DTPA aerosol	37 (1)	No	
^{99m}Tc-MAG3	370 (10)	No	
^{99m}Tc-WBCs	185 (5)	Yes	48 h
^{201}Tl	111 (3)	Yes	96 h

Source: Reprinted by permission of the Society of Nuclear Medicine from Stabin M, Breitz H. Breast milk excretion of radiopharmaceuticals: mechanisms, findings, and radiation dosimetry. J Nucl Med 41:863–873, 2000.

[a] "No" means that interruption of breast-feeding need not be suggested, based on a limit of 1 mSv ED to the infant and these amounts of administered activity, plus other modeling assumptions described in the text. "Yes" means suggestion of interruption with the time intervals noted.

[b] Because of consideration of possible long lived radioactive contaminants.

Abbreviations: DTPA, diethylenetriamine pentaacetic acid; MAA, macroaggregated albumin; EDTA, ethylenediamine tetraacetic acid; DISIDA, disofenin (iminodiacetic acid derivative); HAM, human albumin microspheres; MIBI, methoxyisobutylnitrile; MDP, methylene diphosphonate; PYP, pyrophosphate; RBC, red blood cells; WBC, white blood cells; OIH, orthoiodohippurate; MIBG, metaiodobenzylguanidine; MAG3, mercaptoacetylglycine.

References

1. Traino AC, Martino FDi, Lazzeri M, Stabin MG. Influence of thyroid volume reduction on calculated dose in radioiodine therapy of Graves' hyperthyroidism. Phys Med Biol 45:121–129, 2000.
2. Loevinger R, Budinger T, Watson E. MIRD Primer for Absorbed Dose Calculations. Society of Nuclear Medicine, New York, 1988.
3. Stabin MG, Siegel JA. Physical models and dose factors for use in internal dose assessment. Health Phys 85:294–310, 2003.
4. Stabin MJ, Sparks RB, Crowe E. OLINDA/EXM: the second-generation personal computer software for internal dose assessment in nuclear medicine. J Nucl Med 46:1023–1027, 2005.
5. International Commission on Radiological Protection. Limits for Intakes of Radionuclides by Workers. ICRP Publication 30. Pergamon Press, New York, 1979.
6. International Commission on Radiological Protection. 1990 Recommendations of the International Commission on Radiological Protection. ICRP Publication 60. Pergamon Press, New York, 1991.
7. Poston JW. Application of the effective dose equivalent to nuclear medicine patients. The MIRD Committee. J Nucl Med 34:714–716, 1993.
8. Thomas SR, Stabin MG, Castronovo FP. Radiation-absorbed dose from 201Tl-thallous chloride. J Nucl Med 46:502–508, 2005.
9. International Commission on Radiological Protection. Radiation Dose to Patients from Radiopharmaceuticals. ICRP Publication 53. Pergamon Press, New York, 1988.
10. Cameron JR. A radiation unit for the public. Physics and Society News 20:2, 1991.
11. Weber D, Eckerman K, Dillman LT, Ryman J. MIRD: Radionuclide Data and Decay Schemes. Society of Nuclear Medicine, New York, 1989.
12. International Commission on Radiological Protection. Radionuclide transformations—Energy and Intensity of Emissions. ICRP Publication 38. Pergamon Press, Oxford, 1983.
13. Stabin MG, da Luz CQPL. New decay data for internal and external dose assessment. Health Phys 83:471–475, 2002.
14. Snyder W, Ford M, Warner G. Estimates of specific absorbed fractions for photon sources uniformly distributed in various

organs of a heterogeneous phantom. MIRD Pamphlet No. 5, revised. Society of Nuclear Medicine, New York, 1978.
15. International Commission on Radiological Protection. Report of the Task Group on Reference Man. ICRP Publication 23. Pergamon Press, New York, 1975.
16. International Commission on Radiological Protection: Basic Anatomical and Physiological Data for Use in Radiological Protection. Reference Values. ICRP Publication 89. Pergamon Press, New York, 2003.
17. Snyder W, Ford M, Warner G, Watson S. "S," Absorbed dose per unit cumulated activity for selected radionuclides and organs. MIRD Pamphlet No. 11. Society of Nuclear Medicine, New York, 1975.
18. Cristy M, Eckerman K. Specific absorbed fractions of energy at various ages from internal photons sources. ORNL/TM-8381 V1-V7. Oak Ridge National Laboratory, Oak Ridge, TN, 1987.
19. Stabin M. MIRDOSE—the personal computer software for use in internal dose assessment in nuclear medicine. J Nucl Med 37:538–546, 1996.
20. Stabin M, Watson E, Cristy M, Ryman J, Eckerman K, Davis J, Marshall D, Gehlen K. Mathematical models and specific absorbed fractions of photon energy in the nonpregnant adult female and at the end of each trimester of pregnancy. ORNL Report ORNL/TM-12907. Oak Ridge National Laboratory, Oak Ridge, TN, 1995.
21. Spiers FW, Whitwell JR, Beddoe AH. Calculated dose factors for radiosensitive tissues in bone irradiated by surface-deposited radionuclides, Phys Med Biol 23:481–494, 1978.
22. Eckerman K, Stabin M. Electron absorbed fractions and dose conversion factors for marrow and bone by skeletal regions. Health Phys 78(2):199–214, 2000.
23. Bouchet LG, Bolch WE, Howell RW, Rao DV. S-Values for radionuclides localized within the skeleton. J Nucl Med 41:189–212, 2000.
24. Stabin MG, Eckerman KF, Bolch WE, Bouchet LG, Patton PW. Evolution and status of bone and marrow dose models. Cancer Biother Radiopharm 17:427–434, 2002.
25. Jokisch DW, Patton PW, Inglis BA, Bouchet LG, Rajon DA, Rifkin J, Bolch WE. NMR microscopy of trabecular bone and its role in skeletal dosimetry. Health Phys 75:584–596, 1998.
26. Watson EE, Stabin MG, Davis JL, Eckerman KF. A model of the peritoneal cavity for use in internal dosimetry. J Nucl Med 30:2002–2011, 1989.

27. Stabin MG. A model of the prostate gland for use in internal dosimetry. J Nucl Med 35:516–520, 1994.
28. Bouchet L, Bolch W, Weber D, Atkins H, Poston J Sr. MIRD Pamphlet No. 15: Radionuclide S values in a revised dosimetric model of the adult head and brain. J Nucl Med 40:62S–101S; 1999.
29. Bouchet LG, Bolch WE, Blanco HP, Wessels BW, Siegel JA, Rajon DA, Clairand I, Sgouros G. MIRD Pamphlet No. 19: Absorbed fractions and radionuclide S values for six age-dependent multiregion models of the kidney. J Nucl Med 44:1113–1147, 2003.
30. Stabin MG, Konijnenberg M. Re-evaluation of absorbed fractions for photons and electrons in small spheres. J Nucl Med 41:149–160, 2000.
31. Stabin MG, Stubbs JB, Toohey RE. Radiation dose estimates for radiopharmaceuticals. NUREG/CR-6345, prepared for: U.S. Nuclear Regulatory Commission, U.S. Department of Energy, U.S. Department of Health & Human Services, 1996, 81 pp.
32. International Commission on Radiological Protection. Radiation Dose Estimates for Radiopharmaceuticals. ICRP Publications 53 and 80, with addenda. Pergamon Press, New York, 1983–1991.
33. Snyder W. Estimates of absorbed fraction of energy from photon sources in body organs. In: Medical Radionuclides: Radiation Dose and Effects. USAEC Division of Technical Information Extension, Oak Ridge, TN, 1970, pp. 33–50.
34. Siegel JA, Stabin MG. Mass-scaling of S J values for blood-based estimation of red marrow absorbed dose: the quest for an appropriate method. J Nucl Med 48:253–256, 2007.
35. Russell JR, Stabin MG, Sparks RB, Watson EE. Radiation absorbed dose to the embryo/fetus from radiopharmaceuticals. Health Phys 73:756–769, 1997.
36. Stabin M. Proposed addendum to previously published fetal dose estimate tables for 18F-FDG. J Nucl Med 45:634–635, 2004.
37. Watson EE. Radiation absorbed dose to the human fetal thyroid. In: Fifth International Radiopharmaceutical Dosimetry Symposium. Watson EE, Schlafke-Stelson AT, eds, Oak Ridge Associated Universities, Oak Ridge, TN, 1992, pp. 179–187.
38. Stabin MG, Watson EE, Marcus CS, Salk RD. Radiation dosimetry for the adult female and fetus from iodine-131 administration in hyperthyroidism. J Nucl Med 32:808–813, 1991.
39. Rodriguez M. Development of a kinetic model and calculation of radiation dose estimates for sodium-iodide-131I in athyroid individuals. Master's project, Colorado State University, 1996.

40. Sparks RB, Stabin M. Fetal radiation dose estimates for I-131 sodium iodide in cases where accidental conception occurs after administration. In: Stelson A, Stabin M, Sparks R, eds. Sixth International Radiopharmaceutical Dosimetry Symposium. Oak Ridge Associated Unversities, Oak Ridge, TN, 1999, pp. 360–364.
41. Stabin M, Breitz H. Breast milk excretion of radiopharmaceuticals: mechanisms, findings, and radiation dosimetry. J Nucl Med 41:863–873, 2000.

4
Steps in Dose Calculations

Choice and Application of Standardized Models

In Chapter 3, we reviewed the current state of the art in available models for dose calculations. There are now male and female models for all ages and three models for the pregnant woman. There are also models for rodents as subjects, should dose estimates be of interest in the animals themselves (as opposed to the use of extrapolated animal data for human dosimetry). For most standard dosimetry (package inserts, journal publications), the reference adult (70 kg) is used to represent a standard, typical individual; this allows easy comparison of results from different experimenters. The reference adult was often called "Reference Man," and a 70- to 75-kg individual is still thought to be the best representation of an average adult male across all cultures. This model has traditionally had both male and female organs, so dose to all structures, including female organs, was routinely reported. The slightly smaller (~57 kg) "Reference Woman" has been employed to look at particular issues (see, e.g., "Health Concerns Related to Radiation Exposure of the Female Nuclear Medicine Patient"[1]), and sometimes investigators will present doses for both adult males and females using these models. Individuals in the United States tend to be a bit larger than the world average for adults, because the world average (by definition) includes individuals from many cultures that are below the average and European and American individuals tend to be slightly above the average body size, (not having anything to do with obesity here, just normal stature). Thus the use of the 70-kg model for an average is probably quite reasonable. If one really wishes to show results from both models, this is also reasonable, and, with software automation of dose calculations,[2] this is not difficult. When standard doses are desired for children of different ages, the pediatric model series can also be used, entering the time-activity integrals for the compound and selecting the different phantoms. Very little work has been done in characterizing biokinetics of radiopharmaceuticals in children of different ages, due to logistical difficulties involved with subject enrollment and participation. Typically, the time-activity integrals for adults, taken from imaging studies with human subjects or extrapolated animal data, are simply assumed to apply reasonably well to individuals of all ages. It is logical to suspect that, for some radiopharmaceuticals, radiopharmaceutical kinetics may vary with age, but such information is not generally available.

To calculate dose for an individual subject (patient, research subject, etc.), one should choose the model that is closest to the individual's body mass (not age or gender necessarily, if "hermaphrodite" phantoms are used). Patient-specific adjustments can then be made to the doses for individual organs if desired and if organ masses have been measured, as described in Chapter 3.

Internal Dose Calculations

One way of thinking about internal dose calculations is as a "marriage" of two types of terms: *biology* terms (regarding the biodistribution and retention of the radiopharmaceutical) and *physics* terms (regarding the energy transport and deposition within the body). Although quite time-consuming and difficult, the physics calculations are typically calculated and stored in look-up tables long before any dose calculations are attempted. These are specialized calculations that are carried out by a few laboratories with interest in development of phantoms and models and then made available to the user community. They were previously in printed paper tables but are currently mostly electronic files that are used in automated software programs, such as OLINDA/EXM.[2] Development of the other "half" of the "marriage couple" involves the quantification of data from human or animal studies and treatment of it by a kinetic model, the analysis of which ultimately yields the numbers of disintegrations that have occurred in all significant source organs within the body. Combination of these values with *dose factors* from the standardized phantoms (which give the dose to target regions per disintegration occurring in a source region) then yields the dose estimates that are of interest. We will spend most of the rest of this chapter discussing the gathering of quantitative data for dosimetry.

Input Data: Animal or Human Studies

In Chapter 7, we discuss the U.S. regulatory process for the approval of new medical imaging products. This approval process involves a careful evaluation of the safety and efficacy of such products, using animal and/or human data to produce radiation dose estimates for potential human subjects. As noted in that chapter, any dose estimates based on animal data are to be considered as merely indications of the possibly correct dosimetry, and estimates based on human data are almost always to be preferred, even though these also are fraught with considerable uncertainties. But in many cases, the gathering of animal data is an essential first step in the process of dose evaluation and would be (one would hope) followed by one or more carefully designed and executed human studies that better establish the dose estimates.

Collection of Data

In an important overview document on data gathering and quantification for dosimetry,[3] the authors state the following:

To determine the activity-time profile of the radioactivity in source regions, four questions need to be answered:

1. What regions are source regions?
2. How fast does the radioactivity accumulate in these source regions?
3. How long does the activity remain in the source regions?
4. How much activity is in the source regions?

The first question concerns identification of the source regions, while the second and third questions relate to the appropriate number of measurements to be made in the source regions as well as the timing of these measurements. The fourth question is addressed through quantitative external counting and/or sampling of tissues and excreta.

Each source region must be identified and its uptake and retention of activity as a function of time must be determined. This provides the data required to calculate cumulated activity or residence time in all source regions. Each region exhibiting significant radionuclide uptake should be evaluated directly where possible. The remainder of the body (total body minus the source regions) must usually be considered as a potential source as well. Mathematical models that describe the kinetic processes of a particular agent may be used to predict its behavior in regions where direct measurements are not possible, but where sufficient independent knowledge about the physiology of the region is available to specify its interrelationship with the regions or tissues whose uptake and retention can be measured directly.... The statistical foundation of a data acquisition protocol designed for dosimetry requires that an adequate number of data points be obtained and that the timing of these points be carefully selected. As the number of measurements increases, the confidence in the fit to the data and in the estimates of unknown parameters in the model is improved. As a heuristic or general rule of thumb, at least as many data points should be obtained as the number of initially unknown variables in the mathematical curve-fitting function(s) or in the compartmental model applied to the data set. For example, each exponential term in a multiexponential curve-fitting function requires two data points to be adequately characterized. On the other hand, if it is known a priori that the activity retention in a region can be accurately represented by a monoexponential function, restrictions on sampling times are less stringent as long as enough data points are obtained to

> derive the fitted function. Because of problems inherent in the collection of patient data (e.g., patient motion, loss of specimen, etc), the collection of data above the necessary minimum is advisable.*

These statements are true, whether one is discussing the design of a preclinical study using an animal species to obtain data for extrapolation to humans or the design of a study involving human subjects. Before beginning either type of study, some knowledge of the expected behavior of the radiopharmaceutical will be needed in order to complete a successful study. The first issue of concern is knowledge of which organs and tissues will concentrate the activity to a significant degree: Where will it go? The next issue is an understanding of the expected rates of removal from these tissues: How long will it stay? Most compounds have the tendency to concentrate in a few important tissues and wash out with one or two half-times, with the rest of the activity more or less uniformly distributed in other tissues and being eliminated through the urinary or gastrointestinal excretion pathways. A key point to keep in mind is that the excretory pathways often concentrate as much if not more activity than the organ with the highest concentration of the pharmaceutical due to direct uptake. The study design must include methods for determining which excretion pathways are important and the biokinetics of activity movement through these pathways, or the dosimetry of the compound may be completely misunderstood. For very short-lived radionuclides (e.g., ^{11}C, ^{15}O), some designs will intentionally disregard the issue of excretion, as it is expected that the radionuclide will mostly decay within the body without much excretion occurring. When this assumption is made, for any radiopharmaceutical, the assumption is made

---

*Reprinted with permission of the Society of Nuclear Medicine from Siegel J, Thomas S, Stubbs J, Stabin M, Hays M, Koral K, Robertson J, Howell R, Wessels B, Fisher D, Weber D, Brill A. MIRD Pamphlet No. 16: Techniques for quantitative radiopharmaceutical biodistribution data acquisition and analysis for use in human radiation dose estimates. J Nucl Med 40:37–61, 1999.

that any activity not accounted for in individual organs is uniformly distributed throughout all other organs and tissues of the body and removed only by physical decay. This thus *overestimates* the number of disintegrations in the "remainder of body" compartment but *underestimates* the number of disintegrations that may occur in the urinary bladder or intestines, depending on the excretion pathway(s). The magnitude of this under- and overestimation depends on the degree to which this assumption is not satisfied. For ^{15}O, for example, the assumption is probably very good in most cases. As the physical half-life of the radionuclide increases, the assumption will typically become poorer.

Also of note from the quote cited earlier is the idea that a minimum of two data points per phase of uptake or elimination are always required for important source regions. After one has a reasonable idea of the expected behavior of the radiopharmaceutical, one can then choose when to take samples in order to successfully characterize the biokinetics and ultimately the dosimetry of the compound. For example, radioiodine clears from the body of a euthyroid individual in two phases: the first has a biological half-time of about 6 hours, and the second is much longer, usually quoted as being between 50 and 100 days. Thus, for ^{131}I, the *effective* half-times are around 6 hours and 7 days, respectively. To characterize the clearance of this compound, we would need two points during the first phase of clearance (which is mostly complete by 24 hours postadministration) and then two points during the second, longer phase of clearance. Typical times for data collection might be at 2 to 6 and 18 to 24 hours postadministration for the first phase and then at 48 and 96 hours postadministration to characterize the second phase. Now, logistics may dictate that it is unreasonable to take a second image of a subject 18 hours after administration; if the administration was at, say, 10 AM, this sample would occur at 4 AM the following morning. So the timing of the images can be adjusted to ensure that two points are obtained, spaced as far apart as possible within that phase, but within reasonable times for the patient and staff to be present. Clearly, more points are always helpful in the analysis, as data have inherent uncertainties, and data may be lost at a given time if a patient does not appear or if some equipment malfunction occurs. The analyst would love to have dozens of data points to use in the analysis, but the desire for more data is always balanced against the costs and difficulties of obtaining them.

Extrapolation of Animal Data

In an animal study, the compound of interest may be administered to a number of animals that are then sacrificed at different times, with the activity within the organs estimated by counting: harvesting the organs and counting them in a well counter or other device, or, perhaps, using autoradiography techniques on excised tissue samples, or imaging of the animals (e.g., with a microPET or microSPECT imaging system). The data gathered must be used to predict uptakes in the human from the concentrations seen in the animal tissues (extrapolation). Extrapolation of animal data to humans is by no means an exact science. Crawford and Richmond[4] and Wegst[5] studied some of the strengths and weaknesses of various extrapolation methods proposed in the literature. One method of extrapolating animal data that has been widely applied is the percent kg/g method.[6] In this method, the animal organ data need to be reported as percent of injected activity per gram of tissue, and this information plus knowledge of the animal whole-body weight are employed in the following extrapolation:

$$\left(\frac{\%}{\text{organ}}\right)_{\text{human}} = \left[\left(\frac{\%}{g_{\text{organ}}}\right)_{\text{animal}} \times (\text{kg}_{\text{TBweight}})_{\text{animal}}\right] \times \left(\frac{g_{\text{organ}}}{\text{kg}_{\text{TBweight}}}\right)_{\text{human}}$$

Table 4.1 shows example calculations of data extrapolated from an animal species to the human using this approach.

The animal whole-body weight was 20 g (0.02 kg), and the source organ chosen had a mass of 299 g. The human total

TABLE 4.1. Animal data extrapolation example (mass extrapolation).

	Source organ				
	1 h	3 h	6 h	16 h	24 h
Animal					
%ID/organ	3.79	3.55	2.82	1.02	0.585
(%ID/g)	38.1	36.6	30.8	11.3	5.70
Human					
%ID/organ	3.26	3.12	2.63	0.962	0.486

body weight for the standard adult male of 70 kg was used in the calculations. For example:

$$\frac{38.1\,\%}{\mathrm{g}}\text{(animal)} \times 0.020\,\mathrm{kg} \times \frac{299\,\mathrm{g}}{70\,\mathrm{kg}} = \frac{3.26\,\%}{\text{organ}}\text{(human)}$$

Some researchers have also chosen to perform a transformation of the timescale, to account for the differences in metabolic rate among species of different body mass, which some have suggested can affect the rates at which substances are cleared from the body. One suggested scaling approach is given as:

$$t_{\mathrm{h}} = t_{\mathrm{a}}\left[\frac{m_{\mathrm{h}}}{m_{\mathrm{a}}}\right]^{0.25}$$

where t_a is the time at which a measurement was made in an animal system, t_h is the corresponding time assumed for the human data, and m_a and m_h are the total body masses of the animal species and of the human, respectively. Table 4.2 shows an example case with data extrapolated from an animal species to the human using this timescaling approach.

Here the animal whole-body weight was 200 g (0.2 kg), and again the human total body weight for the standard adult male of 70 kg was used in the calculations. For example:

$$5\text{ min} \times \left[\frac{70\,\mathrm{kg}}{0.2\,\mathrm{kg}}\right]^{0.25} = 22\text{ min}$$

TABLE 4.2. Animal data extrapolation example (time extrapolation).

Animal timescale	5 min	15 min	30 min	60 min	1.5 h
Extrapolated human timescale	22 min	1.1 h	2.2 h	4.3 h	6.5 h

One problematic issue in the area of animal data extrapolation to humans is treatment of activity not accounted for in individual animal organs. Some researchers manage to successfully account for activity in the "carcass" or the rest of the animal body that was not specifically harvested for counting. If the radionuclide is particularly short-lived, this assessment may not be necessary, as one may be able to simply assume that unaccounted-for activity was uniformly distributed in other tissues and removed only by radioactive decay. For many radiopharmaceuticals, however, this will overestimate the number of disintegrations in these "remainder" tissues and underestimate the number of disintegrations in excretory organs like urinary bladder and intestines. An assessment of activity in these regions, via direct counting or analysis of excreta, is usually needed. Such values are usually not extrapolated to humans on a mass basis, but they are assumed to apply directly (i.e., % excreted by the animal = % excreted by the human); a time extrapolation may be applied if desired.

Sparks and Aydogan[7] performed an investigation of the success of animal data extrapolation for a number of radiopharmaceuticals. They reached no concrete conclusions that any particular method was superior to another. They did find, however, that extrapolated animal data tend to *underpredict human organ self-doses*. Figures 4.1 and 4.2 show one example of their results. These figures show the ratio of organ *residence times* (normalized number of disintegrations; see Chapter 2), which is also proportional to organ self-dose, when no extrapolation (Fig. 4.1) or both the time and mass extrapolations (Fig. 4.2) were performed.

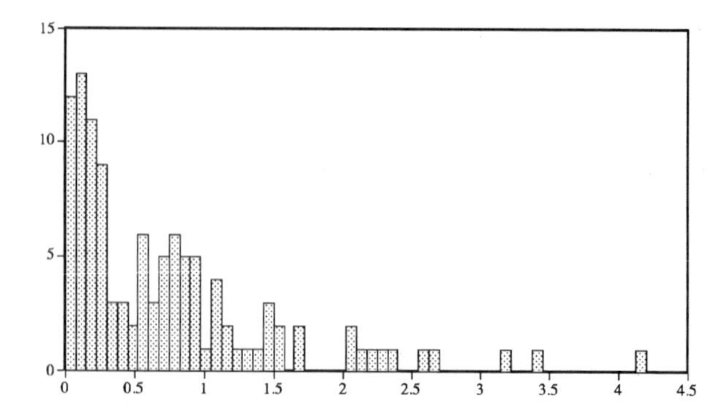

FIGURE 4.1. Frequency distribution of the ratio of the organ residence times found using the raw animal data to the residence times found using data from humans. (Reproduced with permission of the ORAU from Sparks R, Aydogan B. Comparison of the effectiveness of some common animal data scaling techniques in estimating human radiation dose. In: Proceedings of the Sixth International Radiopharmaceutical Dosimetry Symposium. Stelson A, Stabin M, Sparks R, eds, Oak Ridge Institute for Science and Education, Oak Ridge, TN, 1999, pp. 705–716.)

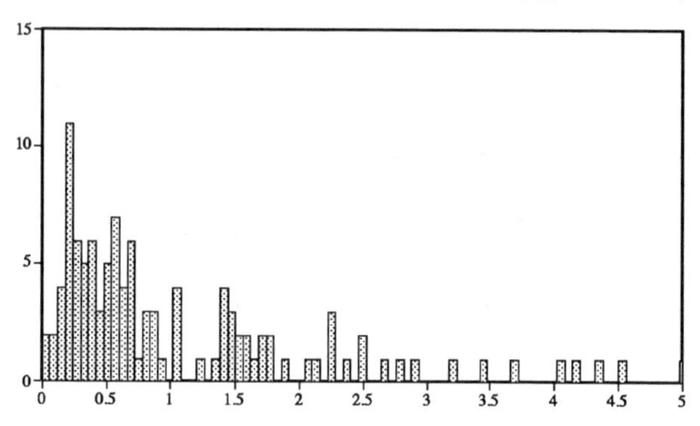

FIGURE 4.2. Frequency distribution of the ratio of the organ residence times found using the time and mass extrapolated animal data to the residence times found using data from humans. (Reproduced with permission of the ORAU from Sparks R, Aydogan B. Comparison of the effectiveness of some common animal data scaling techniques in estimating human radiation dose. In: Proceedings of the Sixth International Radiopharmaceutical Dosimetry Symposium. Stelson A, Stabin M, Sparks R, eds, Oak Ridge Institute for Science and Education, Oak Ridge, TN, 1999, pp. 705–716.)

Image Quantification: Human Data

The external conjugate view counting pair (anterior/posterior) method is the approach most frequently used to obtain quantitative data in human studies for dosimetry. In this method, the source activity A_j is given as[3]:

$$A_j = \sqrt{\frac{I_A I_P}{e^{-\mu_e t}}} \frac{f_j}{C}$$

$$f_j \equiv \frac{(\mu_j t_j/2)}{\sinh(\mu_j t_j/2)}$$

where I_A and I_P are the observed counts over a given time for a given region of interest (ROI) in the anterior and posterior projections (counts/time), t is the patient thickness over the ROI, μ_e is the effective linear attenuation coefficient for the radionuclide, camera, and collimator, C is the system calibration factor (counts/time per unit activity), and the factor f represents a correction for the source region attenuation coefficient (μ_j) and source thickness (t_j) (i.e., source self-attenuation correction). This expression assumes that the views are perfectly collimated (i.e., they are oriented toward each other without offset) and assumes a narrow beam geometry without significant scattered radiation. Corrections for scatter are usually advisable; a number of the proposed methods are described in the following sections.

Corrections for Scattered Radiation

One relatively straightforward correction procedure for scatter compensation involves establishing counting windows on either side of the gamma camera photopeak window such that the area of the two adjacent windows is equal to that of the photopeak; or, if not, the count ratios should be appropriately scaled. The corrected photopeak counts (C_T) are given as:

$$C_T = C_{pp} - F_S(C_{LS} + C_{US})$$

where C_{pp} is the total count recorded within the photopeak window, and C_{LS} and C_{US} are the counts within the lower and upper scatter windows, respectively. The scaling factor (F_S) accounts for the possibility that the total width of the scatter windows (in keV) is not equal to that of the photopeak window and would be unity if they were equal. Thus, subtraction of the adjacent windows is assumed to compensate for the high-energy photon scatter tail upon which the true photopeak events ride. Even if the areas of the scatter windows are equal to that of the photopeak window, use of a scaling factor other than unity may provide the best correction for scatter in a given system with a particular radionuclide. This may be determined by measuring a source of known volume submerged to a realistic depth in a water phantom whose dimensions are similar to that of a human subject.

Corrections for Background Activity

Whenever an ROI is drawn over a source region on a projection image, some counts from the region will have originated from activity in the subject's body that is outside of the identified source region, is scattered radiation from other ROIs, is background radiation, and is due to other sources. Thus, a *background ROI* is drawn over some region of the body that is close to the source ROI and that, in the analyst's opinion, best represents the activity of nearby tissues to the source that will provide the best estimate of a background count rate to be subtracted from the source ROI. As with the scatter correction shown earlier, a scaling factor may be needed to correct the number of counts in the background ROI so that an appropriate correction is made, given the number of pixels in the source and background ROI. Alternately, one may simply subtract the number of counts *per pixel* in the background ROI from the number of counts *per pixel* in the source ROI and then calculate the total number of counts in the source ROI as the corrected number of counts per pixel times the number of pixels. For quantification of counts in the total body or in the check source placed external to the body, an ROI should be drawn away from the subject's body, also away from any "star pattern" streaks that may accompany the source image, but close enough that it captures a typical number of counts per pixel that represent background and scattered radiation within the imaging area close to the subject.

It is important to avoid drawing a background ROI over body structures that may contain a high level of activity (e.g., blood vessels, areas of the skeleton with significant uptake), as this will remove too much background from the source ROI. It is also important to not draw the ROI too far away from the source region in an area of particularly low background, as this may not remove enough background from the source ROI. The choosing of locations and sizes of background ROIs is very difficult to prescribe exactly, and methods vary considerably between investigators, which can result in markedly different results for the final estimates of activity assigned to a source ROI. This process should be carried out with caution and attention to the above points for the best and most reproducible results. The locations of the background ROIs should be documented, perhaps by graphical screen captures if necessary, to enhance the reproducibility of a dosimetric analysis.

Correction for Overlapping Organs and Regions

It is not uncommon for some organs or tumors to have overlapping regions on projection images. The right kidney and liver are frequently partially superimposed on such images, as are the left kidney and spleen, in many subjects. When organ overlap occurs, an estimate of the total activity within a source can be obtained by a number of approximate methods. For paired organs, such as kidneys and lungs, one approach is to simply quantify the activity in one of the organs for which there is no overlap with other organs and double the number of counts in this organ to obtain the total counts in both organs. Another approach is to draw an ROI over the organ region in scans where there is overlap, count the number of pixels, note the average count rate *per pixel*, then use a ROI from another image in which there is no apparent overlap and the whole organ is clearly visible, count the number of pixels in a larger ROI drawn on this image, and then multiply the count rate per pixel from the first image by the number of pixels in the second image. Or, equivalently, take the total number of counts in the first image and multiply by the ratio of the number of pixels in the second to the first image ROIs. If no image can be found in which a significant overlap with another organ does not obscure the organ boundaries, an approximate ROI may need to be drawn just from knowledge of the typical shapes of such organs. This kind of approximation is obviously not ideal, but it may be a necessary approximation.

Obtaining Gamma Camera System Attenuation and Calibration Coefficients

Attenuation Coefficient

The system attenuation coefficient (μ_e) must be measured at some time before (or possibly after) radiopharmaceutical administration in a separate experiment. The basic procedure involves preparation and counting of a source of activity, ideally one whose surface area is greater than that of the source region with the same radionuclide as that to be used for the patient imaging study. As an example, for small regions, fill the bottom of a Petri dish (covered and sealed to prevent possible contamination); for large regions, fill a flood source. A small, point-like source can also be used if necessary. The source should be counted for a fixed time (e.g., 5 minutes) in air, with no intervening attenuating material; then the measurement should be repeated with several different thicknesses of attenuating material of approximately unit density (i.e., 1 g/cm^3) between the source and one of the gamma camera heads. One may obtain the count rates by drawing ROIs encompassing the source region (with correction for background in an adjacent ROI) and then plotting the background-corrected counts in the ROIs versus interposed attenuator thickness.[†] The counts may be fit to an exponentially decreasing function, or the natural logarithm of the counts may be fit to a straight line. In either case, the factor μ_e that best fits the data is the attenuation coefficient to be used in corrections in patient studies.

System Calibration Factor

As with the attenuation coefficient, the system calibration factor, C, must be measured at some time before or after radiopharmaceutical administration in a separate experiment. For this factor, the method is to prepare a standard of known activity of the same radionuclide to be used for admini stration to subjects, usually a few tens of MBq in a suitable container. The exact source strength is not important, as long as sufficient counts are obtained for a consistent evaluation over the course of the study and as long as too many counts are not obtained, which can result in count saturation in the camera. The standard should be counted in air for a fixed time (e.g., 5 minutes) at a source-to-collimator distance that approximates that of the patient midline distance used for the imaging study. The count rate per unit activity (in units of, e.g., cpm/MBq) represents the calibration factor. The collimator count-rate response as a function of the source-to-collimator distance must be known. For parallel-hole collimators, collimator efficiency is invariant; however, for other collimators, such as diverging, converging, and pinhole collimators, the efficiency is dependent upon distance.

It should be noted that, in most cases, the self-attenuation factor f will be approximately equal to unity. Normally, one assumes that the variation in body thickness across individual ROIs is small, and so a single attenuation factor may be used

[†]Another method for acquiring transmission data is use of a transmission scan using a line or flood source.

to calculate the activity for the entire ROI. On the other hand, if the ROI is large and body thickness is thought to vary substantially over the ROI, a pixel-by-pixel calculation may be made. A pixel-by-pixel attenuation calculation can always be made if desired, regardless of this assumption. A conjugate-view measurement is thus made at each of the time points chosen, and the best ROIs for each region are superimposed on the images at each time. Individual organs or tumors may be best visualized at different times of observation. Some regions have most of their uptake early and clear quickly, whereas others may accumulate activity more slowly. Thus, different times may be chosen to draw the best ROI for different regions. The best approach is to have a computer program that allows the ROIs to be independently defined and saved but that then allows them to be linked together and moved together in order to allow the relative locations of all ROIs to be retained when new ROIs are defined, when different patient images reflect slightly different patient placement on the imaging table, or when there is slightly different patient orientation toward the camera heads. Care should be taken to have the patient lie in the same position in all images, as differences in patient orientation toward the camera heads may change the lateral separation between organs. "Bean bag" cushions are widely used for reproducible positioning in radiation oncology. They firm up, conforming snugly to the patient, when a vacuum pump is applied to them. They offer very little additional attenuation.

It is important that, *at all selected imaging time points*, a conjugate-view measurement of a small source of the same radionuclide as that being imaged in the subject be placed within the observed counting region (typically near the feet of the subject). An ROI should be placed over the source and in an appropriate background area outside of the subject; this ROI can also typically be used to correct the "total body" counts within the subject for background. The geometric mean counts of this source, when plotted against time, should decay with the physical half-life of the radionuclide. Any deviation from this might indicate some variability in the counting technique at the different imaging times: shorter or longer scan times, different energy window of acquisition, incorrect collimators used on the camera, or other variations. In general, painfully meticulous attention to detail is necessary in every aspect of quantitative imaging studies. It is sometimes a surprise to a technical staff accustomed to producing images intended only for visual interpretation just how careful one must be to acquire usable, reproducible data for dosimetric purposes.

Kinetic Analysis

Analysis of Kinetic Data

If the investigator has gathered a series of measurements that represent uptake, retention, and/or excretion, the next task is to interpret these measurements in such a way as to derive a kinetic model that can be used to estimate the number of disintegrations occurring in each significant source region in the body. In general, there are three levels of complexity that analysis can take.

Direct Integration

One can directly integrate under the actual measured values by a number of methods. This does not give very much information about your system, but it does allow you to calculate the number of disintegrations rather easily. The most common method used is the *trapezoidal method*, simply approximating the area by a series of trapezoids. An important concern with this method is the calculation of the integrated area under the curve after the last datum. If activity is clearing slowly near the end of the data set, a significant portion of the total decays may be represented by the area under the curve after the last point. A number of approaches may be used to estimate this area. The most conservative is to assume that removal is only by physical decay after the last point; another approach is to calculate the slope of the line using the last two or three points and assume that this slope continues until the retention curve crosses the time axis. No single approach is necessarily right or wrong: a number of approaches may be acceptable under different circumstances. It is generally desirable to overestimate the cumulated activity than to underestimate it, as long as the overestimate is not too severe. The important point is to calculate this area by an appropriate method and to clearly document what was done.

Least-Squares Analysis

An alternative to simple, direct integration of a data set is to attempt to fit curves of a given shape to the data. The curves are represented by mathematical expressions that can be directly integrated. The most common approach is to attempt to characterize a set of data by a series of exponential terms, as many systems are well represented by this form, and exponential terms are easy to integrate. In general, the approach is to minimize the sum of the squared distance of the data points from the fitted curve. The curve will have the form:

$$A(t) = a_1 e^{-b_1 t} + a_2 e^{-b_2 t} + \cdots$$

The method looks at the squared difference between each point and the solution of the fitted curve at that point and minimizes this quantity by taking the partial derivative of this expression with respect to each of the unknowns, a_i and b_i, and setting it equal to zero. Once the ideal estimates of a_i and b_i are obtained, the integral of $A(t)$ from zero to infinity is simply:

$$\int_0^\infty A(t)\,dt = \frac{a_1}{b_1} + \frac{a_2}{b_2} + \ldots$$

If the coefficients a_i are in units of activity, this integral represents cumulated activity: the units of the b_i are time^{-1}. If the coefficients give fractions of the administered activity, then the area represents the normalized cumulated activity (e.g., Bq-h/Bq).

Trapezoidal Method and Least-Squares Analysis Compared

Consider the data set in Table 4.3. We will integrate it by the trapezoidal and least-squares methods.

In the trapezoidal method, each interval is treated separately, and the parts are added:

$A1 = (100 + 72) \cdot 0.5/2 =$ 43 Bq-h
$A2 = (72 + 35) \cdot 0.5/2 =$ 26.75 Bq-h
$A3 = (35 + 24) \cdot 1.0/2 =$ 29.5 Bq-h
$A4 = (24 + 20) \cdot 2.0/2 =$ 44 Bq-h
$A5 = (20 + 15) \cdot 2.0/2 =$ 35 Bq-h
$A6 = (15 + 12) \cdot 4.0/2 =$ 54 Bq-h
Total = 232 Bq-h

In the least-squares analysis, a computer fit of the data yielded the following fit:

$$A(t) = 18.6 \exp(-0.039t) + 81.4 \exp(-1.23t)$$

The cumulated activity for this system, integrating from zero to infinity, then is:

$$\tilde{A} = 18.6/0.039 + 81.4/1.23 = 477 + 66 = 543 \text{ Bq-h}$$

This does not agree well with the estimate from the trapezoidal method. The reason is that in that calculation, we did

TABLE 4.3. Time-activity data for kinetic analysis.

Time (h)	Activity (Bq)
0	100
0.5	72
1	35
2	24
4	20
6	15
10	12

not make an estimate of the area under the curve beyond 10 hours. If we only integrate the analytical expression to 10 hours, the answer is

$$\begin{aligned}\tilde{A} =& 18.6/0.039 \cdot [1 - \exp(-0.039 \cdot 10)] \\ &+ 81.4/1.23 \cdot [1 - \exp(-1.23 \cdot 10)] \\ =& 220\,\text{Bq-h}\end{aligned}$$

This does agree with the trapezoidal estimate. The appropriate calculation to apply to the trapezoidal case is that beyond the last data point activity decreases either with the radioactive half-life or, if it can be estimated reliably, the half-time for the last phase of clearance. In this case, the second phase has a half-time of

$$0.693/0.039 = 17.8\,\text{hours}$$

The area under the curve beyond 10 hours, assuming that this rate continues, is

$$1.443 \cdot 12\,\text{Bq} \cdot 17.8\,\text{hours} = 308\,\text{Bq-h}$$

Adding this value to the previous estimate for the trapezoidal method yields 540 Bq-h, in excellent agreement with the estimate obtained by the least-squares method. Of course, we obtained this second half-time by the least-squares method. If these data were for, say, ^{131}I, and we did not feel that we had a good estimate of this (effective) half-time, we would have to estimate the remaining area as:

$$1.443 \cdot 12\,\text{Bq} \cdot 8.04\,\text{d} \cdot 24\,\text{h/d} = 3340\,\text{Bq-h}$$

This estimate is an order of magnitude higher than the previous estimates and may be quite conservative. Many people, because of the possibility that another, slower clearance phase might exist, will use this assumption even if a least-squares method has been used to fit the existing data. In this case, this highly conservative assumption may unrealistically increase the estimate of the normalized cumulated activity ($\tilde{A}/A_0$). But if a slower component did exist, the assumption that the 17.8-hour clearance rate continued beyond 10 hours could have resulted in a considerable underestimation of the number of disintegrations.

Compartmental Models

The situation frequently arises that you either know quite a bit about the biological system under investigation or you would like to know more about how this system is working. In this case, you can describe the system as a group of compartments linked through transfer rate coefficients. Solving for $\tilde{A}$ of the various compartments involves solving a system of coupled differential equations describing transfer of the tracer between compartments and elimination from the system. The solution to the time-activity curve for each compartment will usually be a sum of exponentials, not obtained by least-squares fitting each compartment separately, but obtained by varying the transfer rate coefficients between compartments until the data are well fit by the model. Computer programs such as SAAM II[8] (Fig. 4.3), Stella,[9] PMod,[10] Simple,[11] and others have been used for these purposes.

Dose Calculations

After one is satisfied with the adequacy of the kinetic data gathered and has performed a kinetic analysis (and thus has estimates of the numbers of disintegrations occurring in each of the important source organs in the body), the final step in the process is to combine the time-activity integrals with the appropriate dose conversion factors, as was outlined in Chapter 2, in order to produce dose calculations for the individual organs that consider contributions from all of the source regions. This can be done by hand or using various mathematical tools, such as spreadsheets, mathematical tools implemented on personal computers or calculators, or software programs specifically designed to calculate

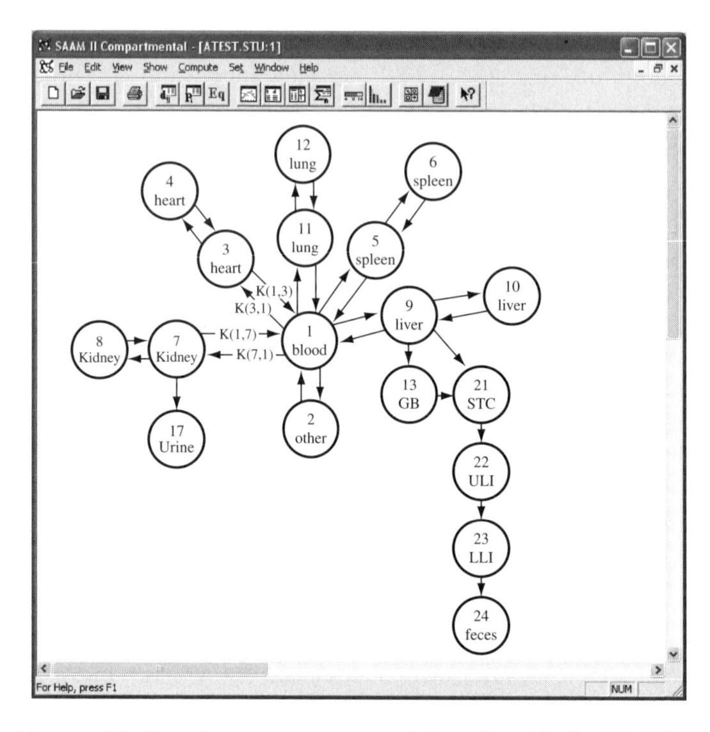

FIGURE 4.3. Sample compartment model, made with the SAAM II program. (Created using SAAM II software: Simulation, Analysis, and Modeling Software for Kinetic Analysis. Software Copyright © 1992-2007 University of Washington, Seattle, WA. All rights reserved. For more information, see http://depts.washington.edu/saam2/.)

internal doses. Given the extent of the calculations needed in most problems, including all necessary corrections (e.g., the "remainder of the body" correction, see Example 4 on page 104), performing the calculations by hand is usually not a very desirable option. The use of some kind of computer or calculator-based automation is usually of interest; the issue then becomes having high confidence in the tool and in the input data in order to trust the output data.

Example calculations, including adjustments often made to some of the standardized models usually available for calculations, will be covered in some of the case studies discussed in Chapter 5. Here in this chapter, we next will overview some basic teaching examples. A mixture of SI units and non-SI (or "antediluvian") units will be employed. SI units are to be highly preferred in all work and publications; sadly, many in the United States stubbornly insist on using antediluvian units although the rest of the scientific community has switched to using the much easier SI system.

Teaching Examples

Example 1: Calculation of S *Value for Average Organ Dose*

We can calculate an S value for liver self-irradiation from ^{99m}Tc by combining the appropriate decay data with calculated absorbed fractions. Table 4.4 shows the decay scheme for ^{99m}Tc from the Brookhaven National Laboratory data source.[12]

At first glance, there appears to be a considerable number of emissions to consider. However, for our purposes, we can consider ^{99m}Tc to have only five emissions: one γ-ray, three X-rays, and a group of nonpenetrating emissions. We can group the nonpenetrating emissions together because they are all multiplied by the same absorbed fraction (1.0), and so, in the sum $\Sigma y_i E_i \phi_i$, we may sum the $y_i E_i$ and multiply the whole sum by $\phi = 1.0$. To calculate the S value for liver irradiating itself, then, we need only to look up the appropriate absorbed fractions for the penetrating emissions (here taken from MIRD Pamphlet No. 5[13] for illustration purposes) and sum over all emissions[‡]:

[‡] *Source*: Data from Snyder W, Ford M, Warner G, Fisher H Jr. MIRD Pamphlet No. 5: Estimates of absorbed fractions for monoenergetic photon sources uniformly distributed in various organs of a heterogeneous phantom. J Nucl Med (Suppl 3):5, 1969.

Emission	y_i	E_i	$k\Sigma y_i E_i$	ϕ	$k\Sigma y_i E_i \phi_\iota$
γ2	0.891	0.1405	0.2666	0.162	0.0432
Kα1 X-ray	0.021	0.0183	0.0008	0.82	0.00067
Kα2 X-ray	0.0402	0.0184	0.0016	0.82	0.00129
Kβ1 X-ray	0.012	0.0206	0.0005	0.78	0.00041
"Nonpenetrating"	—	—	0.0343	1.0	0.0343
				Total = 0.080	

We set k equal to 2.13, which causes the units on the third and fifth columns to be g-rad/μCi-h, given the energy in MeV. The S value is simply the sum of the values in the fifth column divided by the mass of the liver, 1910 g:

$$S(\text{liver} \leftarrow \text{liver}) = 0.080/1800\,\text{g} = 4.18 \times 10^{-5}\,\text{rad}/\mu\text{Ci-h}$$

TABLE 4.4. Decay data for ^{99m}Tc (^{99m}Tc-43 decay mode: IT half-life 6.01 h).

Emission type	Mean energy (MeV)	Frequency
ce-M e-	0.0016	0.7460
Auger-L e-	0.0022	0.1020
Auger-K e-	0.0155	0.0207
ce-K e-	0.1195	0.0880
ce-K e-	0.1216	0.0055
ce-L e-	0.1375	0.0107
ce-L e-	0.1396	0.0017
ce-M e-	0.1400	0.0019
ce-N+ e-	0.1404	0.0004
ce-M e-	0.1421	0.0003
L X-ray	0.0024	0.0048
Kα1 X-ray	0.0183	0.0210
Kα2 X-ray	0.0184	0.0402
Kβ1 X-ray	0.0206	0.0120
γ	0.1405	0.8906
γ	0.1426	0.0002

Source: Adapted with permission from Stabin MG, da Luz CQPL. New decay data for internal and external dose assessment. Health Phys 83:471–475, 2002.

Example 2: Dose to One Organ

Data extrapolated from my own animal study yield the following parameters for a new compound tagged to ^{99m}Tc:

Liver	$f_1 = 0.35$	$T_{e1} = 0.9$ hours
	$f_2 = 0.15$	$T_{e2} = 5.6$ hours
Kidneys	$f = 0.13$	$T_e = 3.4$ hours

where f is the fraction of injected activity. Note that only 63% of the injected activity is accounted for by considering only these two organs. Let's calculate the dose to the liver. If this were a real problem, we would calculate dose to the liver, kidneys, gonads, red marrow, and perhaps a few other organs. We find the following S values in MIRD 11[14]:

$$S(\text{liver} \leftarrow \text{liver}) = 4.6 \times 10^{-5}\,\text{rad}/\mu\text{Ci-h}$$

$$S(\text{liver} \leftarrow \text{kidneys}) = 3.9 \times 10^{-6}\,\text{rad}/\mu\text{Ci-h}$$

(The liver to liver S value is slightly different than we had calculated, as MIRD 11 used slightly different decay data and absorbed fractions.) Assume $A_0 = 1\,\text{mCi} = 1000\,\mu\text{Ci}$; then,

$$\tilde{A}(\text{liver}) = 1.443 \cdot 1000\,\mu\text{Ci}\,(0.35 \cdot 0.9\,\text{h} + 0.15 \cdot 5.6\,\text{h})$$

$$= 1667\,\mu\text{Ci-h}$$

$$\tilde{A}(\text{kidneys}) = 1.443 \cdot 1000\,\mu\text{Ci} \cdot 0.13 \cdot 3.4\,\text{h} = 638\,\mu\text{Ci-h}$$

$$D(\text{liver}) = 1667\,\mu\text{Ci-h}\ 4.6 \times 10^{-5}\,\text{rad}/\mu\text{Ci-h}$$

$$+ 638\,\mu\text{Ci-h} \cdot 3.9 \times 10^{-6}\,\text{rad}/\mu\text{Ci-h}$$

$$D(\text{liver}) = 0.0767\,\text{rad} + 0.0025\,\text{rad} = 0.079\,\text{rad}$$

Note that the liver contributes 97% of its total dose. Dividing by the injected activity, the dose, given these input assumptions, is 0.079 rad/mCi. So, if we redesigned the study to use 3 mCi, the liver absorbed dose would be $3\,\text{mCi} \times 0.079\,\text{rad/mCi} = 0.24\,\text{rad}$.

Let's work the example instead in SI units. We will obtain the same numerical answer for the absorbed dose, as long as we do not run into some round-off error issues:

$$S(\text{liver} \leftarrow \text{liver}) = 3.2 \times 10^{-6}\,\text{mGy/MBq-s}$$

$$S(\text{liver} \leftarrow \text{kidneys}) = 2.9 \times 10^{-7}\,\text{mGy/MBq-s}$$

Assume $A_0 = 37\,\text{MBq}$, equivalent to that above; then,

$$\begin{aligned}\tilde{A}(\text{liver}) =& 1.443 \cdot 37\,\text{MBq}\,(0.35 \cdot 0.9\,\text{h} \\ &+ 0.15 \cdot 5.6\,\text{h}) \cdot 3600\,\text{s/h} \\ =& 2.22 \times 10^{5}\,\text{MBq-s} \\ \tilde{A}(\text{kidneys}) =& 1.443 \cdot 37\,\text{MBq} \cdot 0.13 \cdot 3.4\,\text{h} \cdot 3600\,\text{s/h} \\ =& 8.5 \times 10^{4}\,\text{MBq-s} \\ D(\text{liver}) =& 2.22 \times 10^{5}\,\text{MBq-s} \cdot 3.2 \times 10^{-6}\,\text{mGy/MBq-s} \\ &+ 8.5 \times 10^{4}\,\text{MBq-s} \cdot 2.9 \times 10^{-7}\,\text{mGy/MBq-s}\end{aligned}$$

$$\boxed{\text{D(liver)} = 0.71\,\text{mGy} + 0.025\,\text{mGy} = 0.74\,\text{mGy} = 0.074\,\text{rad}}$$

The answer did not come out quite the same. The dose factors were taken from the OLINDA/EXM code,[2] and the liver to liver value was about 10% different than given in MIRD 11. Both answers are quite reasonable, however, and in good agreement given all of the uncertainties associated with dose calculations.

Example 3: Dose to More than One Organ

A patient is given 1 MBq of ^{99m}Tc: 40% goes to the liver and has a 10-hour biological half-time; the other 60% goes to the spleen and has an infinite biological half-time; no biological removal occurs:

$$\tilde{A}_{\text{liver}} = 4 \times 10^5\,\text{Bq} \times 1.443 \times \frac{6 \times 10\,\text{h}}{6+10} \times \frac{3600\,\text{s}}{\text{h}}$$

$$\tilde{A}_{\text{liver}} = 7.79 \times 10^9\,\text{Bq-s}$$

$$\tilde{A}_{\text{spleen}} = 6 \times 10^5\,\text{Bq} \times 1.443 \times 6\,\text{h} \times \frac{3600\,\text{s}}{\text{h}}$$

$$\tilde{A}_{\text{spleen}} = 1.87 \times 10^{10}\,\text{Bq-s}$$

$$S_{\text{liver}\leftarrow\text{liver}} = 3.08 \times 10^{-6}\,\text{mGy/MBq-s}$$

$$S_{\text{spleen}\leftarrow\text{spleen}} = 2.33 \times 10^{-5}\,\text{mGy/MBq-s}$$

$$S_{\text{spleen}\leftarrow\text{liver}} = 7.2 \times 10^{-8}\,\text{mGy/MBq-s}$$

$$S_{\text{liver}\leftarrow\text{spleen}} = 7.2 \times 10^{-8}\,\text{mGy/MBq-s}$$

$$D_{\text{liver}} = 7.79 \times 10^3\,\text{MBq-s} \times 3.08 \times 10^{-6}\,\text{mGy/MBq-s} +$$
$$1.87 \times 10^4\,\text{MBq-s} \times 7.2 \times 10^{-8}\,\text{mGy/MBq-s} = \underline{\underline{0.025\,\text{mGy}}}$$

$$D_{\text{spleen}} = 7.79 \times 10^3\,\text{MBq-s} \times 7.2 \times 10^{-8}\,\text{mGy/MBq-s} +$$
$$1.87 \times 10^4\,\text{MBq-s} \times 2.33 \times 10^{-5}\,\text{mGy/MBq-s} = \underline{\underline{0.43\,\text{mGy}}}$$

Example 4: Dose to the Fetus, with Remainder of Body Correction

MIRD Dose Estimate Report No. 13[15] gives the numbers of disintegrations for intravenous administration of ^{99m}Tc-MDP as shown in Table 4.5.

If 17 mCi of ^{99m}Tc-MDP has been given to a woman who is 2 weeks pregnant, what is the likely absorbed dose to the fetus? In early pregnancy, up to a few weeks of gestation, the dose to the nongravid uterus is a reasonably good estimate of the fetal dose, because the size and shape of the uterus relative to other organs have not changed substantially. Therefore, we can use S values for these source organs irradiating the uterus:

$$
\begin{aligned}
S(\text{uterus} \leftarrow \text{cortical bone}) &= 5.7 \times 10^{-7}\ \text{rad}/\mu\text{Ci-h} \\
S(\text{uterus} \leftarrow \text{cancellous bone}) &= 5.7 \times 10^{-7}\ \text{rad}/\mu\text{Ci-h} \\
S(\text{uterus} \leftarrow \text{kidneys}) &= 9.4 \times 10^{-7}\ \text{rad}/\mu\text{Ci-h} \\
S(\text{uterus} \leftarrow \text{urinary bladder}) &= 1.6 \times 10^{-5}\ \text{rad}/\mu\text{Ci-h} \\
S(\text{uterus} \leftarrow \text{total body}) &= 2.6 \times 10^{-6}\ \text{rad}/\mu\text{Ci-h}
\end{aligned}
$$

The last S value is not exactly what we need. It is the S value for an organ being irradiated by activity uniformly distributed in the *whole body* (i.e., including bone, kidneys, etc.). The formula for calculating the S value for remainder of the body for a given configuration of other source organs is[16]

$$S(r_k \leftarrow \text{RB}) = S(r_k \leftarrow \text{TB})\left(\frac{m_{\text{TB}}}{m_{\text{RB}}}\right) - \sum_h S(r_k \leftarrow r_h)\left(\frac{m_h}{m_{\text{RB}}}\right)$$

where $S(r_k \leftarrow \text{RB})$ is the S value for "remainder of the body" irradiating target region r_k, $S(r_k \leftarrow \text{TB})$ is the S value for the total body irradiating target region r_k, $S(r_k \leftarrow r_h)$ is the S value for source region h irradiating target region r_k, m_{TB} is the mass of the total body, m_{RB} is the mass of the remainder of the body, i.e., the total body, minus all other source organs used in this problem, and m_h is the mass of source region h.

TABLE 4.5 Numbers of disintegrations for organs in the ^{99m}Tc-MDP model.

Cortical bone	1.36 μCi-h/μCi administered
Cancellous (trabecular) bone	1.36 μCi-h/μCi administered
Kidneys	0.148 μCi-h/μCi administered
Urinary bladder	0.782 μCi-h/μCi administered
Remainder of body	1.64 μCi-h/μCi administered

Source: Data from Weber DA, Makler PT, Jr, Watson EE, Coffey JL, Thomas SR, London J. MIRD Dose Estimate Report No. 13: radiation absorbed dose from ^{99m}Tc labeled bone agents. J Nucl Med 30:1117–1122, 1989.

For this problem, the *S* value for "Remainder of the Body to Uterus" is 2.7×10^{-6} rad/μCi-h (4% higher than that for the total body). The total dose to the uterus is calculated as:

1.36 μCi-h/μCi administered × 5.7×10^{-7} rad/μCi-h = 7.8×10^{-7} rad/μCi
1.36 μCi-h/μCi administered × 5.7×10^{-7} rad/μCi-h = 7.8×10^{-7} rad/μCi
0.148 μCi-h/μCi administered × 9.4×10^{-7} rad/μCi-h = 1.4×10^{-7} rad/μCi
0.782 μCi-h/μCi administered × 1.6×10^{-5} rad/μCi-h = 1.2×10^{-5} rad/μCi
1.64 μCi-h/μCi administered × 2.7×10^{-6} rad/μCi-h = 4.4×10^{-6} rad/μCi
Total = 1.8×10^{-5} rad/μCi

Total dose from incident = 1.8×10^{-5} rad/μCi × 17,000 μCi = 0.30 rad

It would probably be more accurate to use the 57-kg model for the adult female[17] instead of the 70-kg adult male model to calculate this estimate. Using *S* values for the adult female, a dose of 2.3×10^{-5} rad/μCi is estimated, leading to an estimate of the total dose of 0.39 rad.

Example 5: Dose to Several Organs

In MIRD Dose Estimate Report No. 12,[18] the normalized cumulated activity values found for intravenous administration of ^{99m}Tc-DTPA are given as shown in Table 4.6.

TABLE 4.6 Numbers of disintegrations for organs in the ^{99m}Tc-DTPA model.

Kidneys	0.092 MBq-h/MBq administered
Urinary bladder	0.84 MBq-h/MBq administered (2.4-h voiding intervals)
	1.72 MBq-h/MBq administered (4.8-h voiding intervals)
Remainder of body	2.84 MBq-h/MBq administered

Source: Data from Ref. 18.

Let's calculate the absorbed dose to these organs and to the ovaries, testes, and red marrow. For each target organ, then, we will need all of the dose factors for the three source organs. We also have two conditions to the problem: 2.4-hour and 4.8-hour voiding intervals for the urinary bladder. As in the previous example, we will have three contributions to each target organ's total dose for each group of cumulated activity values. An easy way to represent what proves to be a rather substantial amount of math for a simple problem is through the use of matrices. If the set of dose estimates we want is a 2×6 matrix (two sets of dose estimates by six target organs: kidneys, bladder, ovaries, testes, red marrow, and total body), this can be found by multiplication of a 2×3 matrix of cumulated activity values and a 3×6 matrix of dose factors:

$$D = \left(\tilde{A}/A_0\right) \mathrm{DF}$$

$$D = \begin{bmatrix} (\tilde{A}/A_0)_{\mathrm{kid}} & (\tilde{A}/A_0)_{\mathrm{blad1}} & (\tilde{A}/A_0)_{\mathrm{RB}} \\ (\tilde{A}/A_0)_{\mathrm{kid}} & (\tilde{A}/A_0)_{\mathrm{blad2}} & (\tilde{A}/A_0)_{\mathrm{RB}} \end{bmatrix} \times$$

$$\begin{bmatrix} \mathrm{DF(kid \leftarrow kid)} & \mathrm{DF(ov \leftarrow kid)} & \mathrm{DF(mar \leftarrow kid)} & \mathrm{DF(test \leftarrow kid)} & \mathrm{DF(blad \leftarrow kid)} & \mathrm{DF(TB \leftarrow kid)} \\ \mathrm{DF(kid \leftarrow blad)} & \mathrm{DF(ov \leftarrow blad)} & \mathrm{DF(mar \leftarrow blad)} & \mathrm{DF(test \leftarrow blad)} & \mathrm{DF(blad \leftarrow blad)} & \mathrm{DF(TB \leftarrow blad)} \\ \mathrm{DF(kid \leftarrow RB)} & \mathrm{DF(ov \leftarrow RB)} & \mathrm{DF(mar \leftarrow RB)} & \mathrm{DF(test \leftarrow RB)} & \mathrm{DF(blad \leftarrow RB)} & \mathrm{DF(TB \leftarrow RB)} \end{bmatrix}$$

$$D = \begin{bmatrix} 0.092 & 0.842 & 2.84 \\ 0.092 & 1.72 & 2.84 \end{bmatrix} \begin{bmatrix} 1.32 \times 10^{-5} & 7.02 \times 10^{-8} & 1.71 \times 10^{-7} & 3.10 \times 10^{-9} & 1.87 \times 10^{-8} & 1.58 \times 10^{-7} \\ 2.00 \times 10^{-8} & 5.41 \times 10^{-7} & 8.02 \times 10^{-8} & 3.73 \times 10^{-7} & 1.10 \times 10^{-5} & 1.18 \times 10^{-7} \\ 1.54 \times 10^{-7} & 1.81 \times 10^{-7} & 1.34 \times 10^{-7} & 1.28 \times 10^{-7} & 1.67 \times 10^{-7} & 1.39 \times 10^{-7} \end{bmatrix}$$

Actually, the dose factors in the bottom row of the matrix, as taken from the OLINDA/EXM code,[2] are from the total body, not remainder of the body, and need to undergo the adjustment noted above. When this is done, though, we obtain a matrix of dose values (units are mGy/MBq):

$$D = \begin{bmatrix} D_{\text{kid}_1} & D_{\text{ov}_1} & D_{\text{mar}_1} & D_{\text{test}_1} & D_{\text{blad}_1} & D_{\text{TB}_1} \\ D_{\text{kid}_2} & D_{\text{ov}_2} & D_{\text{mar}_2} & D_{\text{test}_2} & D_{\text{blad}_2} & D_{\text{TB}_2} \end{bmatrix}$$

$$D = \begin{bmatrix} 5.5 \times 10^{-3} & 3.5 \times 10^{-3} & 1.7 \times 10^{-3} & 2.4 \times 10^{-3} & 3.5 \times 10^{-2} & 1.8 \times 10^{-3} \\ 5.5 \times 10^{-3} & 5.2 \times 10^{-3} & 1.9 \times 10^{-3} & 3.6 \times 10^{-3} & 7.0 \times 10^{-2} & 2.2 \times 10^{-3} \end{bmatrix}$$

Note from the results the increase in absorbed dose to the bladder, as well as to the gonads, from the increase in the number of disintegrations occurring in the bladder.

Example 6: Correction of Hollow Organ Dose if Source Is in the Wall

Recall that the specific absorbed fraction for electrons in the contents of a hollow organ irradiating the walls is given by:

$$\Phi(\text{wall} \leftarrow \text{contents})_{\text{electrons}} = \frac{1}{2 \times m_{\text{contents}}}$$

Should the source be in the wall instead, the value should be

$$\Phi(\text{wall} \leftarrow \text{wall})_{\text{electrons}} = \frac{1}{m_{\text{wall}}}$$

as is true for the electron self-dose to any organ. The total dose to the organ wall consists of:

$$D_{\text{wall}} = D_{\text{organ 1}} + D_{\text{organ 2}} + \cdots + [D_{\text{photon}}(\text{wall} \leftarrow \text{contents}) + D_{\text{electron}}(\text{wall} \leftarrow \text{contents})]$$

We established earlier that we can assume that $D_{\text{photon}}(\text{wall} \leftarrow \text{contents}) \approx D_{\text{photon}}(\text{wall} \leftarrow \text{wall})$. So if we now assume that the source of radiation is in the wall instead of in the contents, we have only one term to correct. Let's assume that the number of disintegrations occurring in the stomach wall for a ^{11}C agent is 0.0037 MBq-h/MBq administered. The total nonpenetrating component for ^{11}C is 6.17×10^{-14} Gy-kg/Bq-s (1.602×10^{-13} J/MeV $\times 0.386$ MeV/$\beta^+ \times 0.9976$ β^+/dis $\times 1$ dis/Bq-s). The mass of the stomach wall is 158 g, and that of the contents is 260 g. The dose contribution from the contents to the wall would be

$$0.0037 \frac{\text{MBq}-\text{h}}{\text{MBq}} \times 6.17 \times 10^{-14} \frac{\text{Gy kg}}{Bq-s} \frac{1}{2 \times 260\ \text{g}} \times \frac{10^3\ \text{g}}{\text{kg}} \times \frac{3600\ \text{s}}{\text{h}} \times \frac{10^6\ \text{Bq}}{\text{MBq}} \times \frac{10^3 \text{mGy}}{\text{Gy}} = 1.58 \times 10^{-3} \frac{\text{mGy}}{\text{MBq}}$$

Let's assume that the total dose to the stomach wall (from all organs) from an administered ^{11}C compound is 4.6E-03 mGy/MBq, calculated assuming that the source is in the stomach contents (as standard software programs will typically provide, when in reality we think the activity is in the wall). The correct contribution from wall to wall would be

$$0.0037 \frac{\text{MBq}-\text{h}}{\text{MBq}} \times 6.17 \times 10^{-14} \frac{\text{Gy kg}}{\text{Bq}-\text{s}} \frac{1}{158\ \text{g}} \times \frac{10^3\ \text{g}}{\text{kg}} \times \frac{3600\ \text{s}}{\text{h}} \times \frac{10^6\ \text{Bq}}{\text{MBq}} \times \frac{10^3 \text{mGy}}{\text{Gy}} = 5.2 \times 10^{-3} \frac{\text{mGy}}{\text{MBq}}$$

The corrected dose to the stomach wall would be

$$4.6 \times 10^{-3} \frac{\text{mGy}}{\text{MBq}} - 1.58 \times 10^{-3} \frac{\text{mGy}}{\text{MBq}} + 5.2 \times 10^{-3} \frac{\text{mGy}}{\text{MBq}} = 8.2 \times 10^{-3} \frac{\text{mGy}}{\text{MBq}}$$

A shorter path to the same result can be found:

$$4.6 \times 10^{-3} \frac{\text{mGy}}{\text{MBq}} - \left\{ 0.0037 \frac{\text{MBq-h}}{\text{MBq}} \times 6.17 \times 10^{-14} \times 3.6 \times 10^{15} \times \left[\frac{1}{2 \times 260\ \text{g}} - \frac{1}{158\ \text{g}} \right] \right\} = 8.2 \times 10^{-3} \frac{\text{mGy}}{\text{MBq}}$$

Example 7: Effective Dose

One example calculation of the effective dose equivalent was given earlier in this chapter, when the term was first discussed. Another example is worked out here. Assume for a given compound that the liver receives 0.53 mGy, the kidneys receive 0.37 mGy, the ovaries receive 0.19 mGy, the testes receive 0.07 mGy, the red marrow receives 0.42 mGy, the endosteal cells on bone surfaces receive 0.55 mGy, and the thyroid receives 0.05 mGy. Because all of the radiation weighting factors are 1.0, these absorbed doses can be directly converted to equivalent dose (Table 4.7).

The weighting factor for the gonads may be applied to the higher of the values for ovaries or testes. There is a little confusion on this point; ICRP 30[19] used the higher of the two whereas ICRP 53[20] used the average of the two. To use the remainder weighting factor in the ICRP 30 system, one chooses the five organs not assigned an explicit weighting factor that have the highest dose equivalents and assigns them a weighting factor of 0.06. (A different scheme was applied to remainder organs in the ICRP 60 system.) In our example, we only have two to consider. Assign each a factor of 0.06, and ignore the remaining weight of 0.18 (out of 0.30). You could always calculate doses to breast and lung and the other organs and add their contribution, but they will probably be of limited importance. To calculate the H_e, add up the weighted dose equivalents:

Organ	Weighting factor	Dose equivalent (mSv)	Weighted dose equivalent (mSv)
Liver	0.06	0.53	0.0318
Kidneys	0.06	0.37	0.0222
Ovaries	0.25	0.19	0.0475
Red marrow	0.12	0.42	0.0504
Bone surfaces	0.03	0.55	0.0165
Thyroid	0.03	0.05	0.0015
Total (Effective dose equivalent)			0.1699

So we would conclude that the H_e for this compound is 0.17 mSv (0.017 rem or 17 mrem). This suggests that if the whole body were uniformly irradiated to receive 0.17 mSv, the individual would incur the same *additional risk* (of fatal cancer or genetic defects) as from 0.53 mSv to the liver, 0.37 mSv to the kidneys, and so forth.

TABLE 4.7 Equivalent doses converted from absorbed doses.

Organ	Dose equivalent (mSv)
Liver	0.53
Kidneys	0.37
Ovaries	0.19
Testes	0.07
Red marrow	0.42
Bone surfaces	0.55
Thyroid	0.05

Using weighting factors from ICRP 60,[21] let's perform the same calculation[§]:

Organ	Weighting factor	Dose equivalent (mSv)	Weighted dose equivalent (mSv)
Liver	0.05	0.53	0.0265
Kidneys	0.005	0.37	0.00185
Ovaries	0.20	0.19	0.038
Red marrow	0.12	0.42	0.0504
Bone surfaces	0.01	0.55	0.0055
Thyroid	0.05	0.05	0.0025
Total (Effective dose)			0.1248

[§] *Source:* Data from International Commission on Radiological Protection. 1990 Recommendations of the International Commission on Radiological Protection. ICRP Publication 60. Pergamon Press, New York, 1991.

Image-Based, Patient-Individualized Dosimetry

Imaging of patients to obtain anatomic and physiologic information has advanced greatly in the past decade. Anatomic information obtained with magnetic resonance imaging (MRI) or computed tomography (CT) is usually expressed in three dimensions in voxel format, with typical resolutions of the order 1 mm. Similarly, SPECT and PET imaging systems can provide three-dimensional (3D) representation of activity distributions within patients, also in voxel format, with typical resolutions of around 5 to 10 mm. The newest systems now combine CT with PET or SPECT state of the art imaging systems on the same imaging gantry, so that patient anatomy and tracer distribution can be imaged during a single imaging session without the need to move the patient, thus greatly improving and facilitating image registration. Dose calculations based on this approach depend on high-quality SPECT image quantification, which is generally more reliable than planar methods, but it is also more time- and labor-intensive. If done well, however, with appropriate radiation transport methods, one obtains 3D estimates of radiation dose (calculated for each voxel individually, thus not dependent on standard models of the body and its organs). This is a more sophisticated approach, as is used in external beam radiotherapy, and will eventually allow internal dose planning with radiopharmaceuticals to be employed in much the same ways.

The use of a well-supported radiation transport code such as MCNP or EGS4 with knowledge of patient anatomy will result in a significant improvement in the accuracy of dose calculations. Radiation dose calculations for nuclear medicine applications have been mostly relegated to abstract and theoretical calculations, used to establish dosimetry for new agents and to provide reasonable dose estimates to support radiopharmaceutical package inserts and for use in open literature publications. When patients are treated in therapy with radiopharmaceuticals, careful, patient-specific optimization is not performed, as is routine in radiation therapy with external sources of radiation (radiation producing machines, brachytherapy). There are several reasons for this. One involves the limitations on spatial resolution and accuracy of activity quantification with nuclear medicine cameras. Another has to do with the realism and specificity to an individual patient of available body models. The models described above were designed to represent the "reference" adult male and female, children, and so on. Besides using geometric primitives to represent the body and its various organs, only one model is available for any category of individual, so dose estimates calculated using this approach will contain significant uncertainties when applied to any subject, and physicians understandably have low confidence in the use of these results to plan individual subject therapy. Thus, unfortunately for the patients, a "one dose fits all" approach to therapy is usually employed, with significant caution resulting in administration of lower than optimum levels of activity to the majority of subjects (for conservatism and safety). The use of image-based models, not only to develop new "reference" phantoms but also to permit the use of patient-specific models for each therapy patient, is now well developed. Internal dosimetry is thus poised to truly enter into a "Golden Age" in which it will become a more integral part of cancer patient care, much as dosimetry is used in external source radiotherapy. The realism of the newer models is shown in Figure 4.4, with comparison to the form of the existing models developed and implemented in the historical MIRD system.

Several of the efforts to use image data to perform dose calculations, as described earlier, include the 3D-ID code from the Memorial Sloan-Kettering Cancer Center,[22] the SIMDOS code from the University of Lund,[23] the RTDS code at the City of Hope Medical Center,[24] and

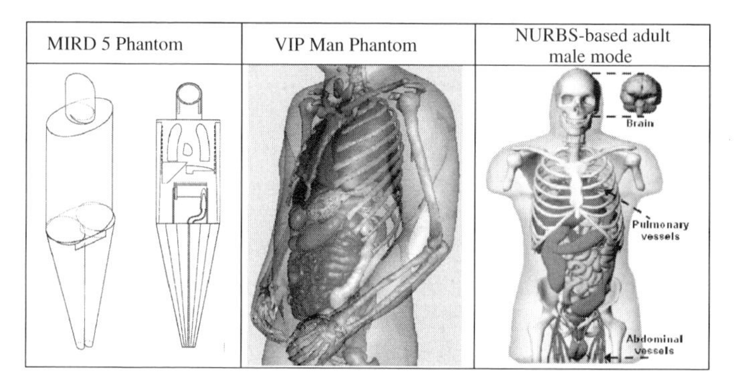

FIGURE 4.4. Comparison of the realism of the traditional MIRD body models with those being used to support current dose modeling efforts. (Frame 1: Reproduced by permission of the Society of Nuclear Medicine from Snyder W, Ford M, Warner G. Estimates of specific absorbed fractions for photon sources uniformly distributed in various organs of a heterogeneous phantom. MIRD Pamphlet No. 5, revised, Society of Nuclear Medicine, New York, 1978. Frame 2: Reproduced with permission from Xu, XG, Chao TC, Bozkurt A. VIP-man: an image-based whole-body adult male model constructed from color photographs of the visible human project for multi-particle Monte Carlo calculations. Health Phys 78:476–486, 2000. Frame 3: Courtesy of Paul Segers, Ph.D., Duke University.)

the DOSE3D code.[25] The code with the most clinical experience to date is the 3D-ID code. These codes either rely on the standard geometric phantoms (MABDose and DOSE3D) or patient-specific voxel phantom data (3D-ID and SIMDOS) and various in-house written routines to perform photon transport. Neither has a particularly robust and well-supported electron transport code, such as is available in EGS[26] or MCNP.[27] The PEREGRINE code[28] has also been proposed for 3D computational dosimetry and treatment planning in radioimmunotherapy.

The usual approach used in these codes is to assume that electron energy is absorbed wherever the electron is first produced. The development and support of electron transport methods is quite complex, as evidenced by ongoing intensive efforts by both the EGS4 and MCNP computer code working groups. It is not reasonable to expect in-house written codes to deal effectively with electron transport. In areas of highly nonuniform activity distribution, such as an organ with multiple tumors evidencing enhanced uptake of an antibody, explicit transport of both photons and electrons is needed to characterize dose distributions adequately.

Investigators at Vanderbilt University have demonstrated the capability for performing radiation transport in voxel phantoms with the MCNP Monte Carlo radiation transport code for internal sources[29,30] in the voxel phantom provided by the group at Yale,[31] and investigators at Rensselaer Polytechnic Institute have demonstrated the capability using the EGS code for external sources in the VIP man voxel phantom.[32,33] Jones[34] reported work performed at the National Radiological Protection Board (NRPB), UK, on an adult male model called NORMAN using MR images of 2-mm × 2-mm resolution and 10-mm slice thickness. Their model was used to estimate organ doses from external photon sources over a range of energies and irradiation geometries. When comparing their calculations with those that used a MIRD-type stylized model, differences in organ doses were found to range from a few percent to more than 100% at photon energies between 10 and 100 keV. Petoussie-Henss et al.[35] reported a family of tomographic models developed from CT images of 2-mm × 2-mm resolution and

8-mm slice thickness. Dose coefficients from external irradiation with these phantoms were substantially different than values derived using the MIRD phantom, suggesting to these authors that the MIRD models do not represent a large proportion of the population well. This is only a partial treatment of what is a rapidly changing and developing field. Developments at present are rapid, with these authors and others contributing new material regularly.

References

1. Stabin MG. Health concerns related to radiation exposure of the female nuclear medicine patient. Env Health Perspect 105(Suppl 6):1403–1409, 1997.
2. Stabin MG, Sparks RB, Crowe E. OLINDA/EXM: the second-generation personal computer software for internal dose assessment in nuclear medicine. J Nucl Med 46:1023–1027, 2005.
3. Siegel J, Thomas S, Stubbs J, Stabin M, Hays M, Koral K, Robertson J, Howell R, Wessels B, Fisher D, Weber D, Brill A. MIRD Pamphlet No 16: Techniques for quantitative radiopharmaceutical biodistribution data acquisition and analysis for use in human radiation dose estimates. J Nucl Med 40:37S–61S, 1999.
4. Crawford DJ, Richmond CR. Epistemological considerations in the extrapolation of metabolic data from non-humans to humans. In: Watson E, Schlafke-Stelson A, Coffey J, Cloutier R, eds. Third International Radiopharmaceutical Dosimetry Symposium. U.S. Department of Health, Education, and Welfare, Washington, DC, 1981, pp. 191–197.
5. Wegst A. Collection and presentation of animal data relating to internally distributed radionuclides. In: Watson E, Schlafke-Stelson A, Coffey J, Cloutier R, eds. Third International Radiopharmaceutical Dosimetry Symposium. U.S. Department of Health, Education, and Welfare, Washington, DC, 1981, pp. 198 203.
6. Kirschner A, Ice R, Beierwaltes W. Radiation dosimetry of 131I-19-iodocholesterol: the pitfalls of using tissue concentration data, the author's reply. J Nucl Med 16:248–249, 1975.
7. Sparks R, Aydogan B. Comparison of the effectiveness of some common animal data scaling techniques in estimating human radiation dose. In: Proceedings of the Sixth International

Radiopharmaceutical Dosimetry Symposium. Stelson A, Stabin M, Sparks R, eds, Oak Ridge Institute for Science and Education, Oak Ridge, TN, 1999, pp. 705–716.
8. SAAM II. Resource for Kinetic Analysis. University of Washington, Seattle, WA. Available at http:// depts.washington.edu/saam2/.
9. Stella. Isee Systems, Lebanon, NH. Available at www.isee systems.com.
10. PMod Technolgoies, Ltd., Zurich, Switzerland.
11. Gambhir SS, Mahoney DK, Turner MS, Wong ATC, Phelps ME. Pet Modeling Tool UCLA. Symbolic Interactive Modelling Package and Learning Environment SIMPLE). A New Easy Method for Computer Modelling. Proc Soc Computer Simulation 173–186, 1996.
12. Stabin MG, da Luz CQPL. New decay data for internal and external dose assessment. Health Phys 83:471–475, 2002.
13. Snyder W, Ford M, Warner G, Fisher H Jr. MIRD Pamphlet No. 5: Estimates of absorbed fractions for monoenergetic photon sources uniformly distributed in various organs of a heterogeneous phantom. J Nucl Med (Suppl 3):5, 1969.
14. Snyder W, Ford M, Warner G, Watson S. MIRD Pamphlet No. 11: "S," absorbed dose per unit cumulated activity for selected radionuclides and organs. Society of Nuclear Medicine, New York, 1975.
15. Weber DA, Makler PT, Jr, Watson EE, Coffey JL, Thomas SR, London J. MIRD Dose Estimate Report No 13: radiation absorbed dose from 99mTc labeled bone agents. J Nucl Med 30:1117–1122, 1989.
16. Cloutier R, Watson E, Rohrer R, Smith E. Calculating the radiation dose to an organ, J Nucl Med 14:53–55, 1973.
17. Stabin M, Watson E, Cristy M, Ryman J, Eckerman K, Davis J, Marshall D, Gehlen K. Mathematical Models and Specific Absorbed Fractions of Photon Energy in the Nonpregnant Adult Female and at the End of Each Trimester of Pregnancy. ORNL Report ORNL/TM 12907, Oak Ridge National Laboratory, Oak Ridge, TN, 1995.
18. Thomas S, Atkins HL, McAfee JG, et al. MIRD Dose Estimate No. 12: Radiation absorbed dose from 99mTc diethylenetriaminepentaacetic Acid (DTPA). J Nucl Med 25:503–505, 1984.
19. International Commission on Radiological Protection. Limits for Intakes of Radionuclides by Workers. ICRP Publication 30. Pergamon Press, New York, 1979.

20. International Commission on Radiological Protection. Radiation Dose Estimates for Radiopharmaceuticals. ICRP Publications 53 and 80, with addenda. Pergamon Press, New York, 1983–1991.
21. International Commission on Radiological Protection. 1990 Recommendations of the International Commission on Radiological Protection. ICRP Publication 60. Pergamon Press, New York, 1991.
22. Kolbert KS, Sgouros G, Scott AM, Bronstein JE, Malane RA, Zhang J, Kalaigian H, McNamara S, Schwartz L, Larson SM. Implementation and evaluation of patient-specific three-dimensional internal dosimetry. J Nucl Med 38:301–308, 1997.
23. Tagesson, Ljungberg M, Strand S. The SIMDOS Monte Carlo Code for the conversion of activity distributions to absorbed dose and dose rate distributions, 416–424, 1996. Proc. 6th International Radiopharmaceutical Dosimetry Symposium. Stelson A, Stabin M, Sparks R, eds, Oak Ridge Associated Universities, Oak Ridge, TN, 1999.
24. Liu A, Williams L, Lopatin G, Yamauchi D, Wong J, Raubitschek A. A radionuclide therapy treatment planning and dose estimation system. J Nucl Med 40:1151–1153, 1999.
25. Clairand I, Ricard M, Gouriou J, Di Paola M, Aubert B. DOSE3D: EGS4 Monte Carlo code-based software for internal radionuclide dosimetry. J Nucl Med 40:1517–1523, 1999.
26. Bielajew A, Rogers D. PRESTA: the parameter reduced electron-step transport algorithm for electron monte carlo transport. Nucl Instrum Methods B18:165–181, 1987.
27. Briesmeister JF, ed. MCNP - A General Monte Carlo N-Particle Transport Code, Version 4C, LA-13709-M. Los Alamos National Laboratory, 2000.
28. Lehmann J, Hartmann Siantar C, Wessol DE, Wemple CA, Nigg D, Cogliati J, Daly T, Descalle MA, Flickinger T, Pletcher D, Denardo G. Monte Carlo treatment planning for molecular targeted radiotherapy within the MINERVA system. Phys Med Biol 50:947–958, 2005.
29. Yoriyaz H, Stabin MG, dos Santos A. Monte Carlo MCNP-4B-based absorbed dose distribution estimates for patient-specific dosimetry. J Nucl Med 42:662–669, 2001.
30. Stabin M, Yoriyaz H. Photon specific absorbed fractions calculated in the trunk of an adult male voxel-based phantom. Health Phys 82:21–44, 2002.

31. Zubal IG, Harrell CR, Smith EO, Rattner Z, Gindi G, Hoffer PB. Computerized 3 dimensional segmented human anatomy. Med Phys 21:299–302, 1994.
32. Chao TC, Bozkurt A, Xu XG. Conversion coefficients based on the VIP-man anatomical model and EGS4-VLSI code for external monoenergetic photons from 10 keV TO 10 MeV. Health Phys 81:163–183, 2001.
33. Xu XG, Chao TC, Bozkurt A . VIP-man: an image-based whole-body adult male model constructed from color photographs of the visible human project for multi-particle Monte Carlo calculations. Health Phys 78:476–486, 2000.
34. Jones DG. A realistic anthropomorphic phantom for calculating specific absorbed fractions of energy deposited from internal gamma emitters. Radiat Prot Dosim 79:411–414, 1998.
35. Petoussi-Henss N, Zankl M, Fill U, Regulla D. The GSF family of voxel phantoms. Phys Med Biol 47:89–106, 2002.

5
Case Studies

In this chapter, I assemble a series of studies of how to analyze real data to obtain usable sets of radiation dose estimates for use in a variety of applications. A few of these data sets are completely manufactured; most are based on real cases that I have analyzed over the past two decades but of course modified in some ways to protect the confidentiality of the original data. These will generally progress from more simple to more complex, and I will try to cover all important aspects of practical dosimetry that I have encountered. For the dose calculations, I will almost always show how to obtain answers using the OLINDA/EXM computer code,[1] not because I wish to promote the use of this code by others, but because, at the time of this writing, it is the only internal dose program that is generally accepted by the dosimetry community (including the Food and Drug Administration). The principles that are illustrated here with this code are expected to apply generally. In most cases, I will use screen-captured images to show as explicitly as possible how each step is performed. My goal with this text is to show in a very practical way how to perform many kinds of dose calculations in nuclear medicine.

Development of New Diagnostic Radiopharmaceuticals

The simplest case that one might encounter is that of a compound that is uniformly distributed throughout the whole body, with a single exponential term describing clearance.

In this case, one would need only to calculate the number of disintegrations occurring in the total body, enter this into the OLINDA/EXM code for "Total Body/Rem Body,"* and calculate dose estimates by choosing a phantom and radionuclide of choice. This is not a very realistic example, however, as a compound with this kind of biokinetic behavior would not be of any great use in nuclear medicine. The simplest *realistic* case that I can cite is that of a short-lived radionuclide, usually a positron emitter, for which we have data for several organs, but no excretion data, and we assume that any activity not accounted for in all of the organs is distributed uniformly throughout all other body tissues and is removed from the body only by radioactive decay.

Case Study 1: Animal Data Set for a ^{11}C-Labeled Emitter, No Excretion

The following data set shows how to implement these simple principles to get useful dosimetry results given these assumptions. Normally in such studies, data may be provided for a large number of tissues, and the first thing to do is to identify the organs for which an extrapolation will be made. Tissues like blood, muscle, skin, connective tissue, extracellular fluid, and others are not employed in the standard body phantoms. Thus, standardized masses are not always available, and extrapolation of results from these tissues is not particularly helpful. There are standard mass values given for blood and blood components, but activity levels in blood are often transient and not useful in establishing the kinetics of activity in the body and its organs.

*This entry value for the code can be a bit confusing. If numbers of disintegrations are entered for any other organs, this entry should be the number of disintegrations for the remainder of the body. In the unusual case in which the only source organ is the total body, this entry should be the number of disintegrations for the total body.

Generally, one looks at the standard, major organs and decides which of them have activity concentrations that are high enough to be worthwhile to study. See Table 5.1 for standard organs usually used in dose calculations. Initially, we can generally ignore organs involved in the excretory pathways (stomach and intestines, gallbladder, and urinary bladder). One does not normally fit time-activity data for these organs; instead, they are usually treated with a standardized excretion kinetic model. These models and specific examples of their use will be described later in this

TABLE 5.1. Standard organs usually used in dose calculations.

Organ
Adrenals
Brain
Breasts
Gallbladder
GI tract
Lower large intestine
Small intestine
Stomach
Upper large intestine
Heart
Kidneys
Liver
Lungs
Ovaries
Pancreas
Skeleton
Active marrow
Cortical bone
Trabecular bone
Spleen
Testes
Thymus
Thyroid
Urinary bladder
Uterus

chapter. The major visceral organs (heart, kidney, liver, lungs, spleen) usually have some significant uptake, and, for some compounds, other organs (e.g., thyroid, brain, adrenals) may have notable concentrations as well. When one views a huge list of reported organ concentrations, it may be difficult to identify where the important uptakes are.

One approach to selecting the important organs for consideration involves a brief study of the reported percent uptake in an organ relative to its percentage contribution to total body mass. In internal dose calculations, we identify the important organs of uptake and assign values of initial activity to them. Activity not assigned is assumed to be distributed throughout all other organs and tissues of the body in accordance with their mass. If a compound is completely uniformly distributed throughout the body in this manner (which, again, would not make for a very useful nuclear medicine agent), each organ would have a percent uptake equal to its percent of total body weight. For example, in the adult male, liver weighs 1910 g, and its percent of total body mass is approximately $1910\,g/73,000\,g = 0.026$, or 2.6%. Thus, if the liver has more than 2.6% uptake at any time, it has a concentration that is above the average for all tissues and should be explicitly treated to determine the number of disintegrations. If no value is above this level, then I would not fit the time-activity data for this organ in many cases. This is by no means a hard and fast rule, but just a guide in making decisions about inclusion of organs for fitting in the kinetic model when there are many measured values. One can certainly always fit all of the data provided and use them in the analysis, but often it is practical to select out five to six important tissues and treat all of the others as "remainder" organs. It is certainly possible to analyze the kinetics of organs that have concentrations below the average for all tissues in the body. Indeed, Cloutier et al.[2] talked of organs that have "deficit" activity, as opposed to those that have high concentrations of activity. In the calculation of organ dose from the "remainder of the body," the dose factor correction method suggested by these authors (and widely accepted in practice since then) will correctly account for organs that have "deficit" activity. This situation is not frequently of major importance (one exception being the brain and an intravenously administered radiopharmaceutical that does not cross the blood-brain barrier), and it is generally a reasonable and conservative assumption to make that these organs be assigned activity as part of the "remainder of the body," in proportion to their mass.

Returning to our example, Table 5.2 gives the data reported for this compound. These data give the %/g in each reported organ multiplied by the animal whole body weight in kg.

The first step in analyzing the data is to convert the animal %-kg/g numbers into human %/organ values (Table 5.3). This is easily done by multiplying each of the values in Table 5.2 by the human organ weights (Table 5.4) and dividing by the human total body mass (73 kg). For example, for the first value in liver:

$$0.01 \times \frac{0.3443 \dfrac{\% - \text{kg}}{\text{g}} 1910\,\text{g}}{73\,\text{kg}} = 0.0901 = 9.01\%$$

Ignoring the blood, muscle, and stomach/intestine data, we find the percentages of total body mass for the remaining organs as given in Table 5.4.

For all of these cases, the maximum uptake in the organ is greater than the percent body weight, so we will fit time-activity curves to all of them and include them in the kinetic model. As there are only three data points, the only model that can be used to fit the data

Table 5.2. Animal organ concentrations (%-kg/g).

Time (h)	Blood	Heart	Lung	Liver	Spleen	Kidney	Small intestine	Large intestine	Stomach	Muscle	Testes	Brain	Bone
0.0833	0.0364	0.12804	0.5712	0.3443	0.3838	0.57964	0.2769	0.1518	0.2622	0.08692	0.0672	0.23478	0.06901
0.75	0.0156	0.02231	0.07392	0.1969	0.06868	0.08686	0.2374	0.0368	0.03312	0.02438	0.0564	0.09009	0.02266
1.5	0.01664	0.02522	0.07104	0.1551	0.05151	0.0946	0.2668	0.0276	0.02392	0.01484	0.0468	0.08463	0.01854

TABLE 5.3. Human organ uptake values (fraction/organ).

Time (h)	Heart	Lung	Liver	Spleen	Kidney	Testes	Brain	Bone
0.0833	0.0055	0.0782	0.0901	0.0096	0.0237	0.00036	0.0457	0.0473
0.75	0.0010	0.0101	0.0515	0.0017	0.0036	0.00030	0.0175	0.0155
1.5	0.0011	0.0097	0.0406	0.0013	0.0039	0.00025	0.0165	0.0127

is that of a single exponential function, that is, $A_{\text{organ}} = A_0 e^{-\lambda t}$, where λ is the biological removal constant (as the data shown above were originally corrected to remove the decay of the ^{11}C). Before beginning the kinetic analysis, it is essential to verify whether or not the data submitted have been corrected for radioactive decay or not. This obviously can make a large difference in the calculated areas under the curve. Fitting one organ at a time, one can obtain the fitted estimates of A_0 and λ. Each function will look something like Figure 5.1.

Each of the N values in Table 5.5 was obtained by the very simple integration:

$$\int_0^\infty \left[A_0 e^{-\lambda_b t}\right] e^{-\lambda_p t} \cdot dt = \int_0^\infty \left[A_0 e^{-(\lambda_b + \lambda_p)t}\right] dt = \frac{A_0}{\lambda_b + \lambda_p}$$

TABLE 5.4. Reference adult organ masses and percent of total body mass.

	Heart	Lung	Liver	Spleen	Kidney	Testes	Brain	Bone
Mass[a]	316	1000	1910	183	299	39.1	1420	5000
Fraction[b]	0.00433	0.01370	0.02616	0.00251	0.00410	0.00054	0.01945	0.06849

[a]Mass of organ in reference adult phantom.
[b]Fraction of total body mass of adult phantom (73000 g) represented by organ.

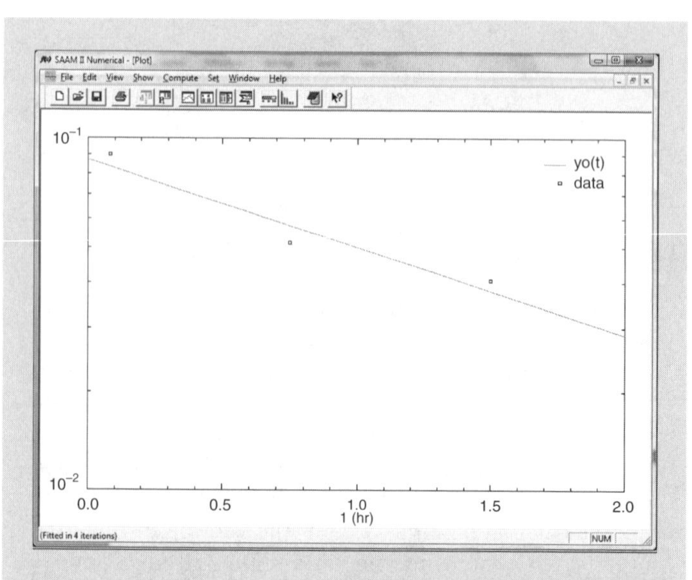

FIGURE 5.1. Fitted function for the liver (human %/organ) data for Case Study 1. Note that the vertical axis is a logarithmic scale, so the single exponential will appear as a straight line with a slope equal to the biological decay constant, λ (assuming that the data points are from decay-corrected data). (Created using SAAM II software: Simulation, Analysis, and Modeling Software for Kinetic Analysis. Software Copyright © 1992–2007 University of Washington, Seattle, WA. All rights reserved. For more information, see http://depts.washington.edu/saam2/.)

For example,

$$N_{\text{heart}} = \frac{0.0022}{0.6448 + 2.04\,\text{h}^{-1}} = 0.000824\,\text{h}$$

Now we have almost all of the data that we need to enter into OLINDA/EXM to obtain dose estimates. The N value for "Bone" must be broken up into cortical and trabecular bone sources. The general rule for this (suggested in ICRP Publication 30)[3] is that for nuclides with physical half-lives shorter than 15 days, one should assume that the activity is distributed

TABLE 5.5. Kinetic model results.

	Heart	Lung	Liver	Spleen	Kidney	Testes	Brain	Bone
A_0	0.0022	0.0736	0.0873	0.0062	0.0206	0.0004	0.0373	0.0385
λ	0.6448	1.4775	0.5866	1.2753	1.3703	0.2781	0.6051	0.8807
N (MBq-h/MBq)	0.000824	0.02107	0.035077	0.001934	0.005808	0.000166	0.014229	0.013706

only on bone surfaces (i.e., not throughout bone volume). The surface area of bone that is involved with actively dividing cells is roughly equal for cortical and trabecular bone,[3] so half of the N value (about 0.00685 MBq-h/MBq) is assigned to both cortical and trabecular bone. A newer ICRP report suggests that this division should be 62% trabecular and 38% cortical.[4] If the nuclide was assumed to be distributed in bone volume, 80% would have been assigned to cortical bone and 20% to trabecular bone. As there is assumed to be no excretion of the activity, the only other value needed is the N for "remainder of body." The N for total body is very easily calculated as just

$$N_{\text{Total Body}} = \frac{1}{\lambda_p} = \frac{1}{(0.693/0.34\,\text{h})} = 0.49\frac{\text{Bq-h}}{\text{Bq}}$$

The value of N for the "remainder of body" is then calculated as:

$$\begin{aligned} N_{\text{Remainder}} &= 0.49 - 0.00082 - 0.021 - 0.035 - 0.0019 \\ &\quad - 0.0058 - 0.000166 - 0.0142 - 0.0137 \\ &= 0.397\frac{\text{Bq-h}}{\text{Bq}} \end{aligned}$$

If these values are entered into the OLINDA/EXM code, the output shown in Table 5.6 is obtained.

One might argue that activity in the stomach and intestines should have been fitted and included in the model, as the concentrations are comparable with or higher than those in other organs. It is not clear whether this activity is in the wall, contents, or both regions of these organs, and other uncertainties about the analysis permits them to be treated reasonably as remainder organs in this case. We will certainly consider cases in which they are treated as distinct organs later.

TABLE 5.6. The OLINDA/EXM program output for values put into the OLINDA/EXM screens for Case Study 1.

	Estimated dose	
Target organ	mGy/MBq	rad/mCi
Adrenals	3.27E-03	1.21E-02
Brain	3.65E-03	1.35E-02
Breasts	2.35E-03	8.69E-03
Gallbladder wall	3.55E-03	1.31E-02
LLI wall	2.91E-03	1.08E-02
Small intestine	3.06E-03	1.13E-02
Stomach wall	3.02E-03	1.12E-02
ULI wall	3.05E-03	1.13E-02
Heart wall	8.40E-03	3.11E-02
Kidneys	6.60E-03	2.44E-02
Liver	6.90E-03	2.55E-02
Lungs	6.60E-03	2.44E-02
Muscle	2.57E-03	9.52E-03
Ovaries	3.02E-03	1.12E-02
Pancreas	3.36E-03	1.24E-02
Red marrow	2.75E-03	1.02E-02
Osteogenic cells	4.60E-03	1.70E-02
Skin	2.10E-03	7.76E-03
Spleen	4.21E-03	1.56E-02
Testes	2.18E-03	8.08E-03
Thymus	2.85E-03	1.05E-02
Thyroid	2.64E-03	9.75E-03
Urinary bladder wall	2.89E-03	1.07E-02
Uterus	3.05E-03	1.13E-02
Total body	2.91E-03	1.08E-02
Effective dose equivalent	4.23E-03	1.57E-02
Effective dose	3.42E-03	1.27E-02
Number of disintegrations in source organs (MBq-h/MBq)		
Brain	1.42E-02	
Heart wall	8.24E-03	
Kidneys	5.80E-03	
Liver	3.50E-02	
Lungs	2.10E-02	
Cortical bone	6.85E-03	
Trabecular bone	6.85E-03	

(Continued)

TABLE 5.6. *(Continued)*

Spleen	1.90E-03
Testes	1.70E-04
Remainder	3.97E-01

Abbreviations: LLI, lower large intestine; ULI, upper large intestine.

Case Study 2: Animal Data Set for an ^{18}F-Labeled Emitter, Urinary Excretion

In this case, we have organ concentration data and urinary excretion data as well. When activity is excreted from the body in the urine, the function that describes it usually consists of one or more exponential terms. Fitting observed activity levels in the urinary bladder is not helpful, as the bladder fills and empties repeatedly, and measurements are too infrequently gathered to characterize this time-activity curve. Material leaving the body is most often governed by first-order processes, which means that the retention (in the body) can be expressed as a function such as $A \times \exp(-\lambda t)$. Therefore, the time-activity curve for the bladder takes the form of $A \times [1 - \exp(-\lambda t)]$, but the curve is periodically interrupted by voiding and goes to zero (or nearly zero) and then begins to accumulate again (Fig. 5.2).

What is needed is a characterization of the values A and λ. (In real situations, there may be more than one term in the equation; but for now, let's just consider one.) In a particularly ingenious derivation, Cloutier and colleagues[5] showed that the number of

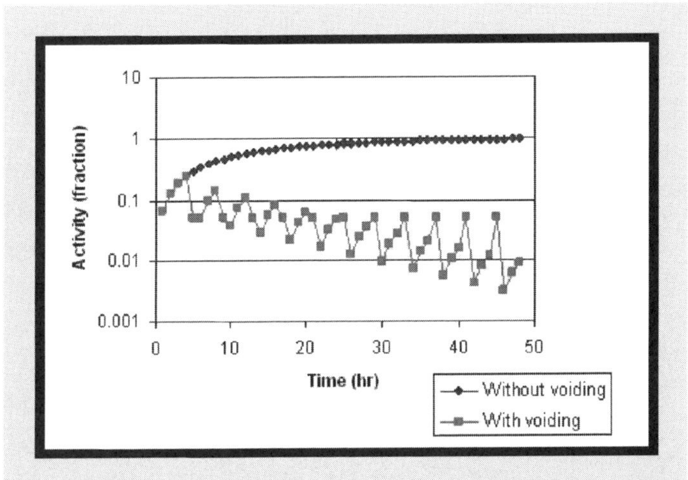

FIGURE 5.2. Urinary bladder time-activity curve, with and without voiding.

disintegrations occurring in the bladder could be given in such cases by a single equation:

$$N = A_0 \sum_i f_i \left[\frac{1-e^{-\lambda_i T}}{\lambda_i} - \frac{1-e^{-(\lambda_i+\lambda_p)T}}{\lambda_i+\lambda_p} \right] \left[\frac{1}{1-e^{-(\lambda_i+\lambda_P)T}} \right]$$

Here, A_0 is the initial activity entering the body, λ_p is the physical decay constant of the radionuclide, λ_i is the biological removal constant for the fraction of activity f_i leaving the body via the urinary pathway, and T is the bladder voiding interval, assumed to be constant. If we have all the activity in the body passing out through the urinary pathway with a 1-hour half-time, for example, our f would be 1.0 and λ would be $0.693/1\,\text{h} = 0.693\,\text{h}^{-1}$. Let's say we have 40% passing out through the GI tract and 60% through the urinary pathway, with two thirds of the urinary clearance having a half-time of 1 hour and one third with a half-time of 10 hours. Then f_1 would be 0.4 and λ_{b1} would be $0.693\,\text{h}^{-1}$, and f_2 would also be 0.2 and λ_{b2} would be $0.0693\,\text{h}^{-1}$.

These parameters are not particularly hard to derive: one must either measure the total body retention or the cumulative urinary excretion and fit a function, either of the form $A \times \exp(-\lambda t)$ (in the former case) or $A \times [1 - \exp(-\lambda t)]$ (in the latter case). Again, the equation may have more than one term, depending on the data observed. If there is GI excretion, this complicates the use of whole-body retention data, unless intestinal activity is somehow excluded from the images. But, in either case, the complication can be overcome by careful data gathering and inspection of the results.

We will now consider a case that has uptake in several organs, activity distributed throughout the "remainder of the body," but also a urinary excretion component. We will not repeat all of the steps shown in the first case study. After the standard data extrapolation, the data shown in Table 5.7 were obtained for human organ uptake.

From these data, the kinetic data shown in Table 5.8 were obtained.

To fit the urinary function, one may fit either the whole-body retention to a function of the form $A \times \exp(-\lambda t)$, or the cumulative urinary excretion to a function of the form $A \times [1 - \exp(-\lambda t)]$. The urinary excretion data shown in Table 5.9 were fit by the second function (Fig. 5.3).

TABLE 5.7. Extrapolated human uptake values.

	% ID/organ human				
Time (h)	Heart	Lungs	Liver	Spleen	Kidneys
0.0833	0.657	1.871	4.643	0.510	2.298
0.75	0.215	0.669	1.577	0.415	0.994
1.5	0.056	0.183	0.423	0.347	0.249

TABLE 5.8. Kinetic parameters for data of Table 5.7.

	Heart	Lungs	Liver	Spleen	Kidneys
A1	0.769	2.191	5.433	0.517	2.780
a1	1.912	1.609	1.540	0.279	1.511
N (MBq-h/MBq)	0.0034	0.0110	0.0283	0.0079	0.0147

TABLE 5.9. Cumulative urinary excretion data.

Time (min)	% excretion
0.50	0.071
1.50	0.143
2.50	0.202
3.50	0.357
4.50	0.559
7.50	1.010
12.50	1.593
17.50	2.163
22.50	2.781
27.50	3.352
32.50	4.041
37.50	4.802
42.50	5.515
47.50	6.300
52.50	.989
57.50	7.583
62.50	8.118
67.50	8.867
72.50	9.449
77.50	10.06
82.50	10.57
87.50	10.86

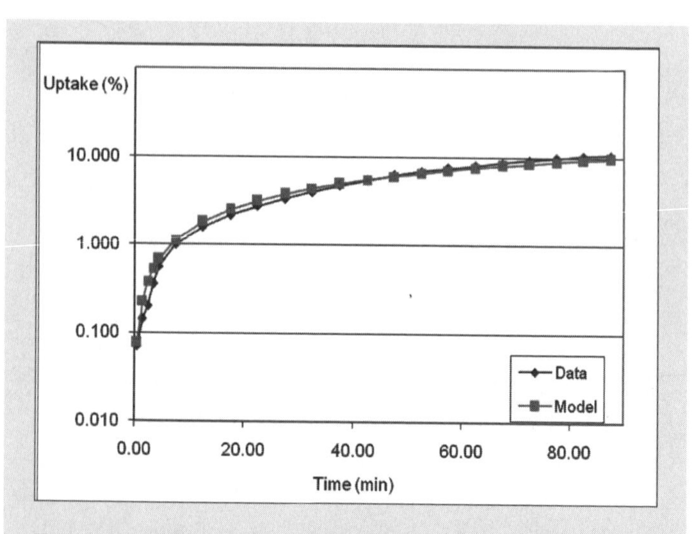

FIGURE 5.3. Fit to data for cumulative urinary excretion, Case Study 2.

The fitted parameters were $A = 20$, $\lambda = 0.0077\ \text{min}^{-1} = 0.462\,\text{h}^{-1}$. The 80% that is not excreted in the urine is assumed to be retained in the total body indefinitely (i.e., removed only by radioactive decay). The calculation for the remainder of the body is

$$N_{\text{TB}} = \frac{0.80}{0.379\,\text{h}^{-1}} + \frac{0.20}{(0.462 + 0.379)\,\text{h}^{-1}} = 2.35\frac{\text{MBq-h}}{\text{MBq}}$$

The remainder of the body calculation will proceed as with the first case:

$$N_{\text{RB}} = 2.35 - 0.0034 - 0.0110 - 0.0283 - 0.0079 - 0.0147$$
$$= 2.29\,\text{MBq-h/MBq}$$

The OLINDA/EXM input screens will look like Figure 5.4A, B (a bladder voiding interval of 4 hours has been chosen for simplicity). The program output is shown in Table 5.10.

Note that, in both of these studies, the activity in "heart" is assigned to "heart wall." This may be a problematic assignment at times. In the case of animal

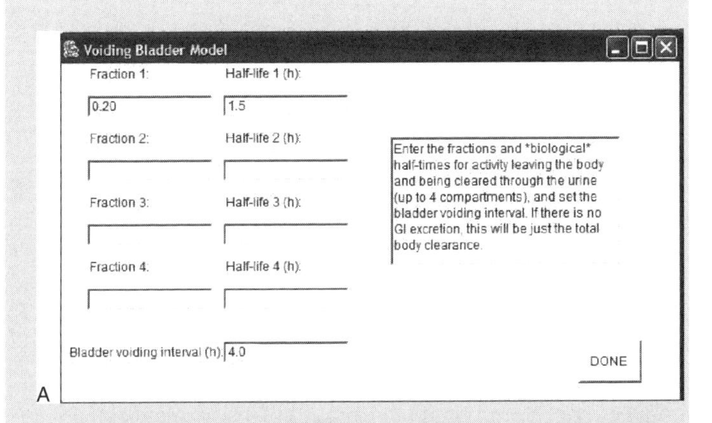

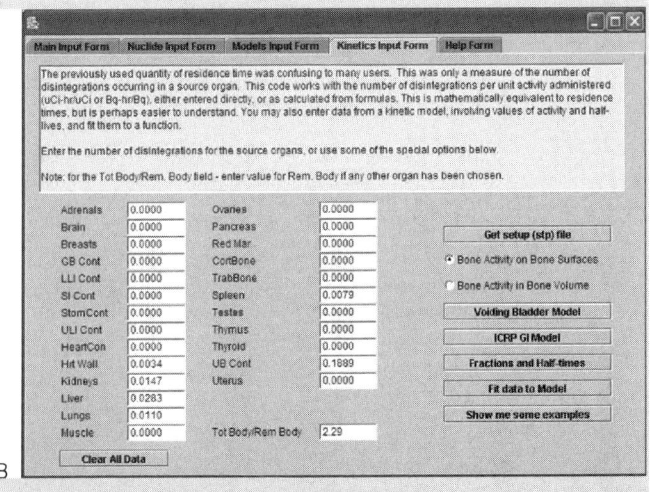

FIGURE 5.4. (A, B) The OLINDA/EXM input screens for voiding bladder model. (Created using The OLINDA/EXM software. Software Copyright © Vanderbilt University.)

studies, the tissue is removed from the animal, and it is reasonable to assume that the contents of the heart are lost before counting begins. In the case of intestines, this is not always as clear. If the contents are not intentionally removed from the walls, they may remain

TABLE 5.10. The OLINDA/EXM program output for the OLINDA/EXM input screens shown in Figure 5.4A, B.

Target organ	Estimated dose	
	mGy/MBq	rad/mCi
Adrenals	1.24E-02	4.58E-02
Brain	9.97E-03	3.69E-02
Breasts	9.05E-03	3.35E-02
Gallbladder wall	1.29E-02	4.77E-02
LLI wall	1.56E-02	5.78E-02
Small intestine	1.43E-02	5.28E-02
Stomach wall	1.25E-02	4.64E-02
ULI wall	1.37E-02	5.07E-02
Heart wall	8.44E-03	3.12E-02
Kidneys	1.52E-02	5.64E-02
Liver	8.19E-03	3.03E-02
Lungs	7.42E-03	2.75E-02
Muscle	1.14E-02	4.23E-02
Ovaries	1.58E-02	5.84E-02
Pancreas	1.32E-02	4.88E-02
Red marrow	1.08E-02	3.99E-02
Osteogenic cells	1.74E-02	6.44E-02
Skin	8.89E-03	3.29E-02
Spleen	1.43E-02	5.30E-02
Testes	1.26E-02	4.66E-02
Thymus	1.12E-02	4.13E-02
Thyroid	1.14E-02	4.23E-02
Urinary bladder wall	1.01E-01	3.75E-01
Uterus	1.90E-02	7.05E-02
Total body	1.15E-02	4.24E-02
Effective dose equivalent	1.83E-02	6.77E-02
Effective dose	1.62E-02	5.98E-02
Number of disintegrations in source organs (MBq-h/MBq)		
Heart wall	3.40E-03	
Kidneys	1.47E-02	
Liver	2.83E-02	
Lungs	1.10E-02	
Spleen	7.90E-03	
Urinary bladder contents	1.89E-01	
Remainder	2.29E+00	

Abbreviations: LLI, lower large intestine; ULI, upper large intestine.

with the contents. In the case of heart, however, it is reasonable (and conservative) to assign all disintegrations to the heart wall. This is conservative because the target organ is the heart wall, and the dose from the wall to the wall is higher than that from the contents to the wall. Note also that the disintegrations in the bladder were not subtracted from the disintegrations in the total body to obtain the number of disintegrations in the remainder of the body. In the case of human studies, patients are instructed to void their bladders before images are taken (with the exception of the early image after injection, in which all counts from the injection should be included in the whole-body image). In the case of animal studies, the activity excreted into the metabolic cages should not be included in the activity estimated in the "whole-body" counting, so this approximation is reasonable.

Case Study 3: Animal Data Set for an ^{18}F-Labeled Emitter, Urinary and Fecal Excretion

Consider the previous case, with the additional consideration that there was an observed maximum uptake of 25% of the administered activity in the intestines. Fitting of observed activity in the intestines is generally not highly reliable. It is difficult to quantify the activity in the intestines: in the case of animal studies, the issue is the separation of the wall and contents activity; in the case of human studies, the issue is the unique identification of the various regions of the intestines and the elimination of overlap with activity in other regions of the abdomen (e.g., liver, kidneys, spleen). The 10% excretion in the urine remains the same. The calculation for the remainder of the body changes to the following:

$$N_{\mathrm{TB}} = \frac{0.55}{0.379\,\mathrm{h}^{-1}} + \frac{0.20}{(0.462+0.379)\,\mathrm{h}^{-1}} = 1.69\frac{\mathrm{MBq\text{-}h}}{\mathrm{MBq}}$$

We introduce 25% (a fraction of 0.25) into the ICRP 30 GI tract model, assumed to enter at the small intestine. This results in assigned numbers of disintegrations in the intestinal organs of:

Small intestine	0.398 MBq-h/MBq
Upper large intestine	0.218 MBq-h/MBq
Lower large intestine	0.040 MBq-h/MBq

The remainder of the body calculation changes to:

$$N_{\mathrm{RB}} = 1.69 - 0.0034 - 0.0110 - 0.0283 - 0.0079 - 0.0147$$
$$= 1.63\,\mathrm{MBq\text{-}h/MBq}$$

Note that the disintegrations in the intestines were not subtracted here. The calculation of the number of disintegrations in the "total body" did not include these disintegrations, so they did not need to be subtracted to obtain the correct number to assign to remainder tissues. If the number of disintegrations were, by contrast, based on human image data, disintegrations in the intestines would have been in the whole-body region of interest and should have been subtracted. (Disintegrations in the urinary bladder, however, would not have been, as patients normally void their bladders before scans are taken.) The rest of the calculation proceeds as in Case Study 2. The resultant absorbed doses are shown in Table 5.11.

TABLE 5.11. The resultant absorbed doses for Case Study 3.

Target organ	Estimated dose	
	mGy/MBq	rad/mCi
Adrenals	1.04E-02	3.86E-02
Brain	7.15E-03	2.64E-02
Breasts	6.76E-03	2.50E-02
Gallbladder wall	1.75E-02	6.48E-02
LLI wall	4.21E-02	1.56E-01
Small intestine	9.60E-02	3.55E-01
Stomach wall	1.30E-02	4.79E-02
ULI wall	1.07E-01	3.97E-01
Heart wall	7.19E-03	2.66E-02
Kidneys	1.69E-02	6.26E-02
Liver	9.19E-03	3.40E-02
Lungs	6.27E-03	2.32E-02
Muscle	1.03E-02	3.80E-02
Ovaries	2.58E-02	9.55E-02
Pancreas	1.20E-02	4.46E-02
Red marrow	1.08E-02	4.00E-02
Osteogenic cells	1.37E-02	5.08E-02
Skin	7.08E-03	2.62E-02
Spleen	1.43E-02	5.30E-02
Testes	1.02E-02	3.77E-02
Thymus	8.22E-03	3.04E-02
Thyroid	8.23E-03	3.05E-02
Urinary bladder wall	1.01E-01	3.75E-01
Uterus	2.52E-02	9.31E-02
Total body	1.14E-02	4.22E-02
Effective dose equivalent	3.25E-02	1.20E-01
Effective dose (mSv/MBq)	2.35E-02	8.69E-02
Number of disintegrations in source organs (MBq-h/MBq)		
LLI	3.99E-02	
Small intestine	3.98E-01	
ULI	2.18E-01	
Heart wall	3.40E-03	
Kidneys	1.47E-02	
Liver	2.83E-02	
Lungs	1.10E-02	
Spleen	7.90E-03	
Urinary bladder contents	1.89E-01	
Remainder	1.63E+00	

Abbreviations: LLI, lower large intestine; ULI, upper large intestine.

A situation that arises at times involves the idea that the activity in the intestines was really in the walls of the intestine and not in the contents, as the OLINDA/EXM code assumes. The important issue here is that, if the activity is in the walls and not in the contents, the electron dose from wall to wall has a specific absorbed fraction of $1/m_w$ and not $1/(2m_c)$ (m_w is the mass of the organ wall, and m_c is the mass of the contents). The photon component of the dose will be approximately the same for a source in the wall or contents. So a fairly simple manipulation of the computer-provided result can make this correction. The steps are

1. For the organ, separate what portion of the organ's total dose was contributed by the activity assigned to its contents. The OLINDA/EXM code provides the percent contributions to individual organ dose totals. For this case, the code reports that small intestine contributed 81.8% of its own dose, or $0.818 \times 0.096\,\text{mGy/MBq} = 0.0785\,\text{mGy/MBq}$.
2. Now separate the photon and electron components of this self-dose. For this, you need the nuclide delta function for all of the electrons. For ^{18}F, this only involves the positron and one Auger electron.[6] The electron self-dose is

$$\frac{\Delta}{2 \times m_C} = \frac{3.86 \times 10^{-14}\,\dfrac{\text{Gy-kg}}{\text{Bq-s}}}{2 \times 0.423\,\text{kg}} = 4.56 \times 10^{-14}\,\frac{\text{Gy}}{\text{Bq-s}}$$

$$4.56 \times 10^{-14}\,\frac{\text{Gy}}{\text{Bq-s}} \times 0.398\frac{\text{MBq-h}}{\text{MBq}} \times \frac{3600\,\text{s}}{\text{h}} \times \frac{1000\,\text{mGy}}{\text{Gy}}$$

$$\times \frac{10^6\,\text{Bq}}{\text{MBq}} = 6.54 \times 10^{-2}\,\frac{\text{mGy}}{\text{MBq}}$$

3. The photon contribution is thus $0.0785 - 0.0654 = 0.0132\,\text{mGy/MBq}$. Add this contribution to the correct electron contribution for wall irradiating wall,

and you have the corrected dose, assuming activity was really in the wall and not the contents:

$$\frac{\Delta}{m_W} = \frac{3.86 \times 10^{-14}\,\frac{\text{Gy-kg}}{\text{Bq-s}}}{0.677\,\text{kg}} = 5.7 \times 10^{-14}\,\frac{\text{Gy}}{\text{Bq-s}}$$

$$5.7 \times 10^{-14}\,\frac{\text{Gy}}{\text{Bq-s}} \times 0.398\,\frac{\text{MBq-h}}{\text{MBq}} \times \frac{3600\,\text{s}}{\text{h}} \times \frac{1000\,\text{mGy}}{\text{Gy}}$$

$$\times \frac{10^6\,\text{Bq}}{\text{MBq}} = 8.17 \times 10^{-2}\,\frac{\text{mGy}}{\text{MBq}}$$

$$8.17 \times 10^{-2}\,\frac{\text{mGy}}{\text{MBq}} + 1.32 \times 10^{-2}\,\frac{\text{mGy}}{\text{MBq}} = \underline{\underline{9.48 \times 10^{-2}\,\frac{\text{mGy}}{\text{MBq}}}}$$

So the *self-dose* changes from 0.0785 mGy/MBq to 0.0948 mGy/MBq, and the *total dose* is now 0.0948mGy/MBq + (1 − 0.818) × 0.096 mGy/MBq = 0.112 mGy/MBq.

Case Study 4: Human Data from Gamma Camera Images

Moving now to cases involving estimates based on human image data, we will consider first the most common approach, which is the use of planar image data. Figure 5.5 shows anterior and posterior images of a subject given a radiopharmaceutical. The posterior image has been inverted left-to-right so that the organs are in the same locations as the anterior image, and a region of interest (ROI) has been drawn around the left

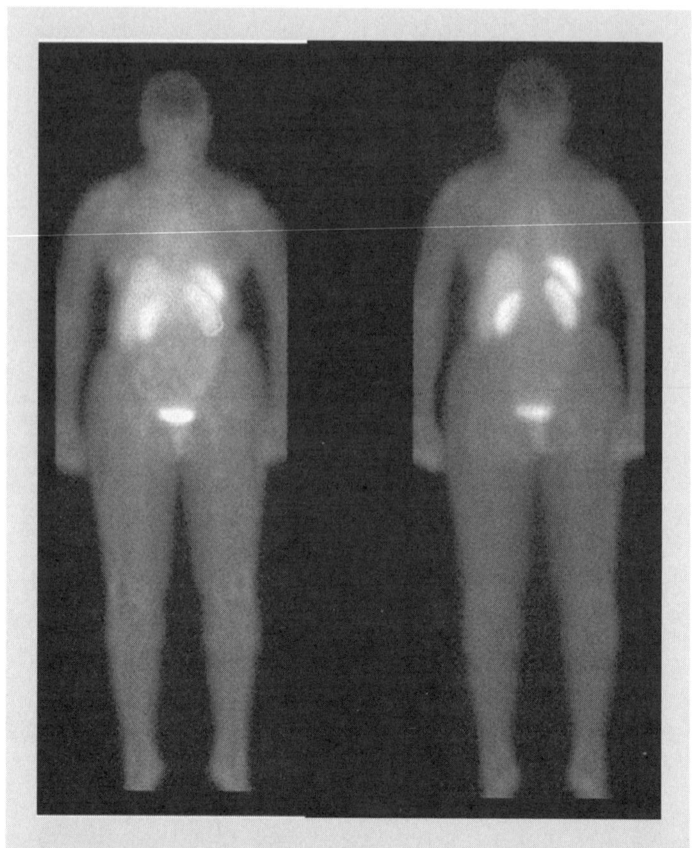

FIGURE 5.5. Anterior (left) and posterior (right) images of a subject given a radiopharmaceutical. The posterior image has been inverted so that the organs are in the same locations as the anterior image, and a region of interest (ROI) has been drawn around the left kidney on the anterior image. (Reproduced from http://www.rad.kumc.edu/nucmed/clinical/siadh.htm. Permission granted by the University of Kansas Hospital, Division of Nuclear Medicine, Kansas City, KS.)

kidney on the anterior image. For this subject, the right kidney is also well defined, and it appears that a ROI could be drawn over this organ as well, maintaining separation from the liver. The spleen has significant uptake of the pharmaceutical also and is very clearly defined. The liver is apparent, but it has a somewhat unusual shape. Nonetheless, a ROI could be drawn over the liver, and counts could be extracted. There appears to be some minor uptake in the chest, which might be blood pool activity in the heart or perhaps uptake in the lungs. Drawing ROIs over the lungs from these images would be difficult, and this might be done better using ^{57}Co transmission images, if available. Interestingly, a deficit of activity may be noted in the brain, suggesting that this compound does not cross the blood-brain barrier. One could draw a ROI over the brain area and estimate a number of disintegrations for this region, which would be lower than predicted by distributing the "remainder body" disintegrations uniformly everywhere. This would be more correct, but some investigators may not wish to go to this trouble, and they would just let the brain dose be calculated conservatively as being similar to that of other remainder tissues. A ROI should *not* be drawn over the bladder region. As noted earlier, the time-activity curve for bladder is far too complicated to be fit by a finite number of data points that may be sampled over time—a typical human study for dosimetry will have only perhaps three to seven time points—and the number of disintegrations in the bladder should be obtained by characterization of the urinary excretion or whole-body retention, exactly in the same manner as for the animal studies discussed earlier. If urine samples are obtained, one may fit the urinary excretion to a function of the form of $A \times [1 - \exp(-\lambda t)]$, and retention (in the body) can be expressed as a function such as $A \times \exp(-\lambda t)$. When the A and λ terms are defined (again, possibly involving more than one component, so A_1, λ_1, A_2, λ_2, etc.), the dynamic bladder model may be used exactly as was done above.

An important consideration in the extracting of counts from planar images is the drawing of appropriate *background* ROIs for organs and the whole body. Drawing the ROI to subtract counts from outside the body is usually relatively easy (at least in comparison with drawing ROIs for internal body structures). Background ROIs are usually just small circular or elliptical regions that are placed in an area that seems "reasonable" in representing counts that are underlying the image at all points where the ROI for the organ or whole body is drawn. In the case of the whole-body background ROI, this basically represents room background and scattered radiation from the subject. Any reasonable placement of the background ROI in the image field will give an estimate of this background count rate. Usually, these counts are very low compared with the total counts per pixel inside the body, and so precise placement of this region is not terribly important. At very late times after injection, the count rates in the body may be quite low and more comparable with the background count rates, but, at this point, the contributions to dose are also less important. The drawing of ROIs inside of the body, however, is far more problematic. Looking at Figure 5.5, where would one reasonably draw a background ROI for the liver? for the kidneys? for the lungs? For the organs in the lower abdomen, one might draw a region adjacent to these organs, but still within the body, clearly, but avoiding any regions with enhanced activity that do not really represent the body background. One should generally avoid drawing a region in the center of the abdomen, as it is easy to pick up counts occurring in major blood vessels, in the marrow in the pelvis or spine, or in the intestines. In Figure 5.5, it is not entirely clear, but there appears to be some uptake in the intestines, and activity in the femoral arteries is also seen. Just above the liver to the right, there appears as well to be a small focus of activity, perhaps a tumor or some specific uptake in the lung; one would certainly not draw a region here. Different investigators will choose different placements for background ROIs in the body, and it is at times impressive to note the differences in the ultimate estimation of organ uptake that can occur. Drawing background ROIs is more of an art than a science, and the investigator is simply warned to approach this procedure with care. It is advisable when one is uncertain to ask the opinion of a knowledgeable colleague in this area.

There are a number of computer programs that have been developed to extract numerical values of counts from ROIs drawn over planar image data. No single program stands out as superior to all others or as most commonly employed by many users. The ImageJ code[7] is notable, as it is freeware. It is developed by investigators at the National Institutes of Health (NIH) and distributed via Internet download. Once a code is chosen, the images at various times can be loaded into the code, ROIs can be drawn, and counts can be extracted. As noted in Chapter 4, what we wish to do now is implement the external conjugate view method, in which the source activity A_j is given as:

$$\mathrm{A_j} = \sqrt{\frac{I_\mathrm{A} I_\mathrm{P}}{e^{-\mu_\mathrm{e} t}}} \left(\frac{f_\mathrm{j}}{C}\right)$$

$$f_\mathrm{j} \equiv \frac{(\mu_\mathrm{j} t_\mathrm{j}/2)}{\sinh\,(\mu_\mathrm{j} t_\mathrm{j}/2)}$$

where I_A and I_P are the observed counts over a given time for a given ROI in the anterior and posterior projections (counts/time), t is the patient thickness over the ROI, μ_e is the effective linear attenuation coefficient for the radionuclide, camera, and collimator, C is the system calibration factor (counts/time per unit activity), and the factor f represents a correction for the source region attenuation coefficient (μ_j) and source thickness (t_j) (i.e., source self-attenuation correction).

The first step in analyzing a series of patient images, taken over several days, is to calculate the times at which the data were taken, for use in the kinetic analysis. This is easily done with any spreadsheet program by just subtracting one time from another, as in Table 5.12.

Even though the study design may have called for images at 1, 4, 24, 48, and 72 hours, it is not sufficient to use these as the times for fitting the data. The actual, precise times postadministration must be used. Obviously, depending on the radionuclide, this will be perhaps a minor or major influence on the results. For example, for an ^{131}I-labeled agent, the difference between 72 hours and 76.27 hours is only about 1.5%, whereas for a ^{99m}Tc-labeled agent, it is nearly 64%! Then the counts from the various ROIs need to be analyzed in the time sequence. Many tools can perform this service; simple mathematical spreadsheets are one

Table 5.12. Human imaging acquisition times.

Time of administration:1/17/2005, 12:08

Scan times and dates		Time postadministration (h)
1/17/2005	13:35	1.45
1/17/2005	15:39	3.52
1/18/2005	16:09	28.02
1/19/2005	15:40	51.53
1/20/2005	16:24	76.27

such tool. Assuming that data for a liver ROI, with a suitable background ROI, were taken over a series of pairs of images at the times above, the information in Table 5.13 might be found.

This process must be repeated for all organs, including the total body. This gives us the terms I_A and I_P, as noted earlier. As discussed in Chapter 4, the system calibration factor C (counts/time per unit activity) will have been obtained previously by counting a source of known activity with the same energy windows and collimators that were used to obtain these data. The patient thickness over the ROI (t) will be obtained by copying the ROI for the liver over the same spot on two ^{57}Co flood field transmission scans: one with the patient and bed in place between the ^{57}Co source, and one with only the bed in place. The effective thickness is given as:

$$t = \frac{-\ln \frac{C_2}{C_1}}{\mu_{Co}}$$

where C_2 are the counts in the ROI with the patient in place, C_1 are the counts without the patient present, and μ_{Co} is the effective linear attenuation coefficient for ^{57}Co, as determined via an attenuation study, also as outlined in Chapter 4. The factor μ_e is the attenuation coefficient for the radionuclide of interest, also obtained from an attenuation study. As noted in Chapter 4, the self-attenuation correction may be performed, but rarely does this have a strong impact on the final results. If data were gathered in energy windows adjacent to the peak for scatter correction purposes, they will be directly subtracted from the values in the right hand column of Table 5.13, using the appropriate weighting factors, as determined by each investigator as being appropriate for her system. If the width of the scatter window(s) is equal to that of the

TABLE 5.13. Human gamma camera data (liver).

Time	Organ no. pixels	Organ counts/pixel	View	Background no. pixels	Background counts/pixel	Net counts/pixel	Net counts
1.45	2812	31.67	Anterior	540	8.00	23.7	66,560
1.45	2812	26.75	Posterior	540	6.53	20.2	56,855
3.52	2812	28.64	Anterior	540	8.50	20.1	56,641
3.52	2812	23.51	Posterior	540	6.71	16.8	47,241
28.02	2812	11.47	Anterior	540	4.57	6.9	19,403
28.02	2812	8.14	Posterior	540	3.61	4.5	12,757
51.53	2812	4.50	Anterior	540	1.96	2.5	7155
51.53	2812	3.62	Posterior	540	1.63	2.0	5599
76.27	2812	1.81	Anterior	540	0.85	1.0	2725
76.27	2812	1.57	Posterior	540	0.68	0.9	2484

Table 5.14. Time-activity data for liver.

Time	Activity (counts)	Activity (fraction of administered activity)
1.45	167,471	0.0875
3.52	140,824	0.0736
28.02	42,830	0.0224
51.53	17,231	0.00900
76.27	7082	0.00370

primary photopeak window, the scaling factor may be just 1.0. The factor may vary considerably, however, and it is best determined for each nuclide, camera system, and collimator set-up and optimized. Instead of doing this as a ROI averaged value, the determination may also be made on a *pixel-by-pixel basis*[8] and the contributions summed at the end. Eventually, then, the scatter and attenuation corrected counts will be related to the amount of activity originally administered and the activity in an organ expressed as the fraction of administered activity (Table 5.14; Fig. 5.6).

This process is repeated for all organs and the total body. Procedures for handling urinary and/or fecal excretion are exactly the same as those described above for animal studies. One other potentially problematic organ to estimate activity for, however, is the red marrow. Two approaches exist for quantifying uptake in this region when it is thought to be significant. One is to quantify the amount of activity in the blood as a function of time and assume that the uptake in marrow can be related to the uptake in blood:

$$[A_{\text{marrow}}] = [A_{\text{blood}}] \times \text{RMBLR}$$

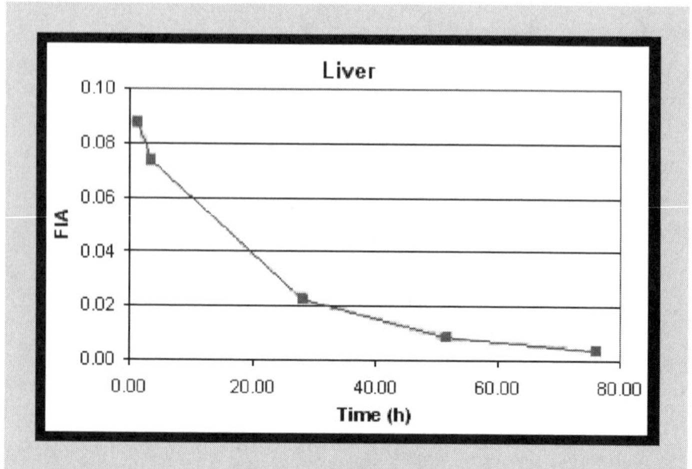

FIGURE 5.6. Curve fit to liver time-activity data.

where $[A_{\text{marrow}}]$ is the concentration of compound (assumed in this publication to be a monoclonal antibody) in the marrow, $[A_{\text{blood}}]$ is the concentration of the agent in the blood or serum, and RMBLR is the red marrow to blood cumulated activity ratio. One expression of this, used by many, was proposed by Sgouros[9]:

$$[A_{\text{marrow}}] = [A_{\text{blood}}]\frac{\text{RMECFF}}{(1-\text{HCT})}$$

Here, RMECFF is the vascular and extracellular fluid (ECF) volume in the marrow, and HCT is the patient hematocrit. The "working" value for the RMECFF was suggested to be 0.19. Other authors[10] have adapted this method to other agents, assuming different values for the RMECFF. The other method is to draw a ROI over some region of marrow that can be clearly visualized apart from other structures in the body (e.g., lumbar spine, appendicular skeleton) and extract counts as for any other organ.[11] The difficulty in this method is that the fraction of total marrow associated with

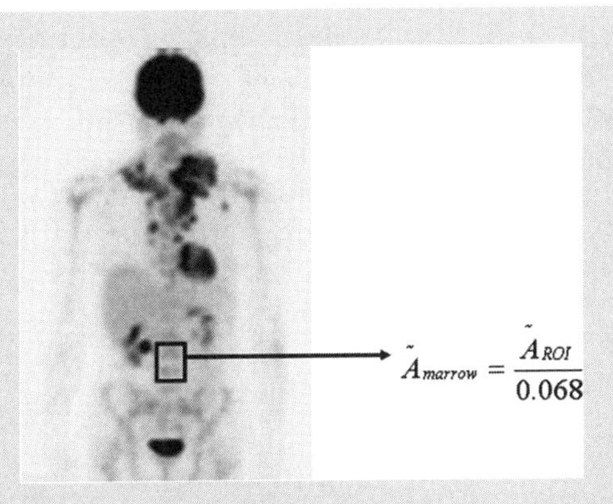

FIGURE 5.7. Measurement of counts over marrow region and extrapolation to total marrow activity. (Reproduced from http://www.mcg.edu/radscape/Case Studies/Spring%202005/Krista/Case.htm. Courtesy of the Department of Nuclear Medicine, Medical College of Georgia.)

this region must be assumed, generally from published standard values.[12,13] In Figure 5.7, we have drawn ROIs over a region in the abdomen that appears to represent marrow uptake. This region was drawn to represent the marrow in two lumbar vertebrae (L4 and L5). According to Cristy,[12] the marrow in these two vertebrae comprises 6.8% of the total marrow in the skeleton. Thus, the number of disintegrations found in this region should be divided by 0.068 to obtain an estimate of the number of disintegrations in the whole marrow.

Case Study 5: Kinetic Analysis

The next step in the process is to perform a kinetic analysis of these data to determine the number of disintegrations occurring in all source regions. Let's consider the liver time-activity curve we developed earlier from the human data set and see how the estimates of number of disintegrations might be different using different methods to integrate the data set. First, we will use the direct integration technique, using a trapezoidal method. For the first time interval, the area under the curve is a triangle, going from zero activity at time zero to the value of 0.08747 at time 1.45 hours:

$$A = 0.5 \times 0.08747 \frac{\text{MBq in organ}}{\text{MBq administered}} \times 1.45\ \text{h}$$
$$= 0.0634 \frac{\text{MBq-h}}{\text{MBq}}$$

For all of the other regions, the area is one half of the difference in the fraction of injected activity (FIA) values, multiplied by the difference in times, for example:

$$A = 0.5 \times (0.08747 + 0.07356) \frac{\text{MBq in organ}}{\text{MBq administered}}$$
$$\times (3.52 - 1.45)\ \text{h} = 0.1664 \frac{\text{MBq-h}}{\text{MBq}}$$

The sum of the areas to this point is 1.93 MBq-h/MBq (Table 5.15). Assuming that this is an ^{131}I-labeled agent, the most conservative thing to do is to assume only physical decay after the last data point. Doing this, we would then add $1.443 \times 0.0037 \times 8\,\text{d} \times 24\,\text{h/d} = 1.03$ MBq-h/MBq to the area so far, thus calculating a total area of 2.96 MBq-h/MBq. A second approach taken by some is to estimate the slope of the last two to three data points and assume that this slope is maintained after the end of the data. Fitting the last three points to a straight line, one obtains the function

TABLE 5.15. Fitted time-activity data using the trapezoidal method.

Time	Activity (FIA)	Area
1.45	0.08747	0.0634
3.52	0.07356	0.1664
28.02	0.02237	1.1751
51.53	0.00900	0.3689
76.27	0.00370	0.1570

$$y = 3.85\text{E-}04x + 3.17\text{E-}02$$

This will cross the y axis at $x = 82.34$ hours. So the area from 76.27 hours to 82.34 hours is

$$0.5 \times 0.0037 \times (82.34 - 76.27) = 0.0112 \frac{\text{MBq-h}}{\text{MBq}}$$

This gives a very different estimate of the total area under the curve of $1.93 + 0.0112 = 1.94$ MBq-h/MBq.

Now, what if instead we fit the data to a single exponential function? These data are very well fit by a function $A(t) = 0.0842 \times e^{-0.0422t}$. Integrating this function from 0 to 76.27 hours, we obtain:

$$\frac{0.0842 \frac{\text{MBq}}{\text{MBq administered}}}{0.0422\ \text{h}^{-1}}(1 - e^{-0.0422 \times 76.27}) = 1.915 \frac{\text{MBq-h}}{\text{MBq}}$$

This is in very nice agreement with the trapezoidal method estimate. If we assume that the function continues to infinity, the integral is just:

$$\frac{0.0842 \frac{\text{MBq}}{\text{MBq administered}}}{0.0422\ h^{-1}} = 1.995 \frac{\text{MBq-h}}{\text{MBq}}$$

Note that inherent in these calculations is the assumption that the data presented included the decay of ^{131}I. If the data had been corrected for the decay of

^{131}I instead, this calculation would have been:

$$\frac{0.0842 \frac{\text{MBq}}{\text{MBq administered}}}{(0.0422 + 0.0036)\,\text{h}^{-1}} = 1.839 \frac{\text{MBq-h}}{\text{MBq}}$$

In this case, before performing the trapezoidal integration above, all of the points would have needed to be multiplied by $e^{-0.0036t}$, where t is the time at which the data point was observed.

Finally, these data were fit in a multicompartmental model with the SAAM II software.[14] The model had the form shown in Figure 5.8.

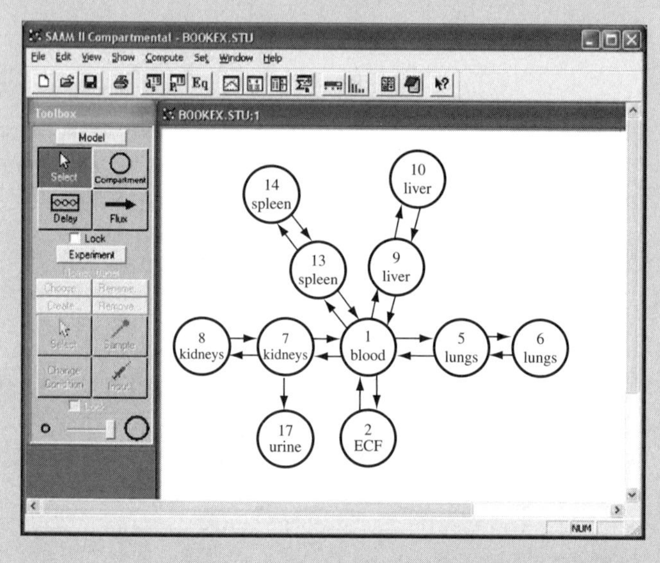

FIGURE 5.8. Kinetic model, Case Study 5. (Created using SAAM II software: Simulation, Analysis, and Modeling Software for Kinetic Analysis. Software Copyright © 1992–2007 University of Washington, Seattle, WA. All rights reserved. For more information, see http://depts.washington.edu/saam2/.)

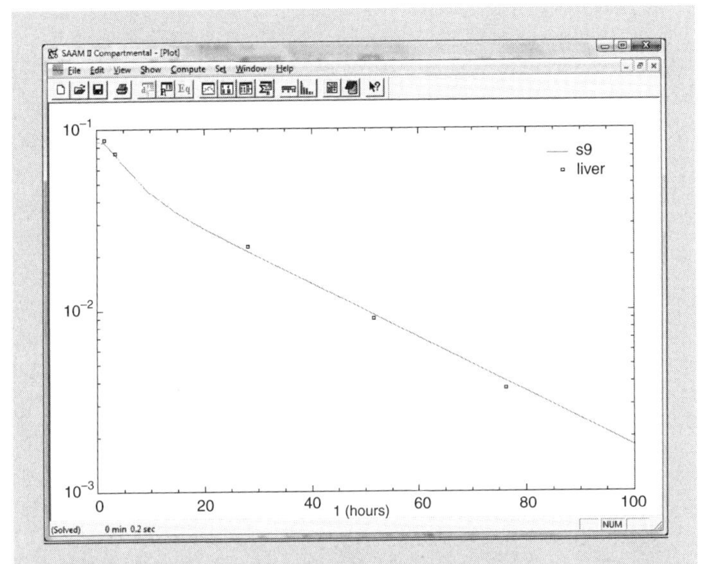

FIGURE 5.9. Compartment model fit to liver data. (Created using SAAM II software: Simulation, Analysis, and Modeling Software for Kinetic Analysis. Software Copyright © 1992–2007 University of Washington, Seattle, WA. All rights reserved. For more information, see http://depts.washington.edu/saam2/.)

Data for spleen, kidneys, lungs, and remainder of the body were also fit. The fit for the liver data was quite good as shown in Figure 5.9.

The estimated area under the curve was 1.81 MBq-h/MBq, in basically good agreement with the estimate from the regression analysis or the second estimate using the trapezoidal method.

Case Study 6: Case with Ascites

One of the specialized organs developed some years back for internal dosimetry is a model of the peritoneal cavity.[15] Some agents are used to evaluate the level of ascites observed in patients; animal studies as well have been developed for these agents. Calculation of the number of disintegrations in the peritoneal cavity may be performed via image data analysis or sampling of the ascites fluid. Once these values are ascertained, the integration progresses as described earlier. To perform the dosimetry for the subject, one may employ the dose conversion factors for this peritoneal cavity model presented by Stabin and Siegel[16] and employed in the OLINDA/EXM software.[1] The only complication here is that this special model is not integrated with the rest of the software dose calculations for the standard body organs. Thus, one must perform the calculations for the standard organs and for the peritoneal cavity contributions separately and then manually add the results at the end. Because the dose estimates are given per unit activity administered, as long as the sum of the disintegrations used in the two parts of the calculation are based on a unit administration of activity, the results may just be added directly to give the total dose from all sources. Consider an example in which we have a ^{99m}Tc-labeled agent with the following numbers of disintegrations in the source regions:

$N_{liver} = 0.257$ MBq-h/MBq
$N_{lungs} = 0.129$ MBq-h/MBq
$N_{peritoneal\ cavity} = 0.420$ MBq-h/MBq
$N_{urinary\ bladder} = 0.112$ MBq-h/MBq
$N_{remainder} = 2.07$ MBq-h/MBq

The doses from the standard organs are given by the OLINDA/EXM code as shown in Table 5.16.

The program gives the doses from the peritoneal cavity component as listed in Table 5.17.

TABLE 5.16. Radiation dose estimates for ^{99m}Tc agent.

Target organ	Estimated dose	
	mGy/MBq	rad/mCi
Adrenals	1.69E-03	6.27E-03
Brain	9.37E-04	3.47E-03
Breasts	9.02E-04	3.34E-03
Gallbladder wall	2.03E-03	7.53E-03
LLI wall	1.56E-03	5.79E-03
Small intestine	1.54E-03	5.69E-03
Stomach wall	1.38E-03	5.10E-03
ULI wall	1.53E-03	5.65E-03
Heart wall	1.57E-03	5.80E-03
Kidneys	1.43E-03	5.30E-03
Liver	3.57E-03	1.32E-02
Lungs	2.53E-03	9.35E-03
Muscle	1.14E-03	4.23E-03
Ovaries	1.63E-03	6.03E-03
Pancreas	1.73E-03	6.39E-03
Red marrow	1.17E-03	4.31E-03
Osteogenic cells	3.16E-03	1.17E-02
Skin	7.55E-04	2.80E-03
Spleen	1.30E-03	4.82E-03
Testes	1.13E-03	4.19E-03
Thymus	1.25E-03	4.62E-03
Thyroid	1.16E-03	4.29E-03
Urinary bladder wall	5.62E-03	2.08E-02
Uterus	1.91E-03	7.06E-03
Total body	1.29E-03	4.77E-03

Abbreviations: LLI, lower large intestine; ULI, upper large intestine.

As noted above, these results are directly additive, as the sum of the disintegrations in the two parts of the calculation represents 100% of the disintegrations occurring for the product. A slight logistic problem is encountered, as the organs used in the peritoneal cavity model are in some cases slightly different and

TABLE 5.17. Radiation dose estimates for ^{99m}Tc agent in peritoneal cavity.

Doses from nuclide: ^{99m}Tc in peritoneal cavity	
Target organ	Dose (mGy/MBq)
Adrenals	8.89E-04
Brain	4.19E-07
Breasts	5.58E-05
LLI wall	2.57E-04
Small intestine	1.03E-03
Stomach wall	6.32E-04
ULI wall	9.07E-04
Heart wall	3.87E-04
Kidneys	6.70E-04
Liver	5.67E-04
Lungs	1.59E-04
Ovaries	6.15E-04
Pancreas	2.39E-03
Muscle	2.21E-04
Skeleton	1.78E-04
Active marrow	2.78E-04
Bone surfaces	2.71E-04
Skin	4.26E-05
Spleen	3.64E-04
Testes	4.25E-05
Thymus	5.26E-05
Thyroid	1.35E-06
Urinary bladder wall	3.98E-04
Uterus	2.69E-03
Whole body	2.40E-04

Abbreviations: LLI, lower large intestine; ULI, upper large intestine.

in some cases presented in a slightly different order than those given in the OLINDA/EXM output. This is not difficult to resolve in any standard spreadsheet program, and the calculation is as shown in Table 5.18.

TABLE 5.18. Adding of doses from organ and peritoneal cavity dose contributions.

Target organ	Dose (mGy/MBq)
Adrenals	1.69E-03
Brain	9.37E-04
Breasts	9.02E-04
Gallbladder wall	2.03E-03
LLI wall	1.56E-03
Small intestine	1.54E-03
Stomach wall	1.38E-03
ULI wall	1.53E-03
Heart wall	1.57E-03
Kidneys	1.43E-03
Liver	3.57E-03
Lungs	2.53E-03
Muscle	1.14E-03
Ovaries	1.63E-03
Pancreas	1.73E-03
Red marrow	1.17E-03
Osteogenic cells	3.16E-03
Skin	7.55E-04
Spleen	1.30E-03
Testes	1.13E-03
Thymus	1.25E-03
Thyroid	1.16E-03
Urinary bladder wall	5.62E-03
Uterus	1.91E-03
Total body	1.29E-03

+

Target organ	Dose (mGy/MBq)
Adrenals	8.89E-04
Brain	4.19E-07
Breasts	5.58E-05
LLI wall	2.57E-04
Small intestine	1.03E-03
Stomach wall	6.32E-04
ULI wall	9.07E-04
Heart wall	3.87E-04
Kidneys	6.70E-04
Liver	5.67E-04
Lungs	1.59E-04
Muscle	2.21E-04
Ovaries	6.15E-04
Pancreas	2.39E-03
Active marrow	2.78E-04
Bone surfaces	2.71E-04
Skin	4.26E-05
Spleen	3.64E-04
Testes	4.25E-05
Thymus	5.26E-05
Thyroid	1.35E-06
Urinary bladder wall	3.98E-04
Uterus	2.69E-03
Whole body	2.40E-04

=

Target organ	Dose (mGy/MBq)
Adrenals	2.58E-03
Brain	9.37E-04
Breasts	9.58E-04
Gallbladder wall	2.03E-03
LLI wall	1.82E-03
Small intestine	2.57E-03
Stomach wall	2.01E-03
ULI wall	2.44E-03
Heart wall	1.96E-03
Kidneys	2.10E-03
Liver	4.14E-03
Lungs	2.69E-03
Muscle	1.36E-03
Ovaries	2.25E-03
Pancreas	4.12E-03
Red marrow	1.45E-03
Osteogenic cells	3.43E-03
Skin	7.98E-04
Spleen	1.66E-03
Testes	1.17E-03
Thymus	1.30E-03
Thyroid	1.16E-03
Urinary bladder wall	6.02E-03
Uterus	4.60E-03
Total body	1.53E-03

Abbreviations: LLI, lower large intestine; ULI, upper large intestine.

Case Study 7: Radiocontaminants in Radiopharmaceutical Product

Following Case Study 6, we consider now a commonly encountered calculation, that of radiopharmaceutical products that have known radioactive contaminants. In many cases, but not necessarily all, the physical half-life of these products is longer than that of the primary radionuclide, and the dosimetry characteristics are less desirable: for example, many of the radiocontaminant isotopes of the element involved in the radiopharmaceutical (e.g., radioactive iodine species in iodine radiopharmaceuticals, such as ^{124}I, ^{125}I, ^{126}I, and ^{130}I in ^{123}I products, or ^{200}Tl and ^{202}Tl products in ^{201}Tl chloride). The presence of the contaminants is usually expressed in terms of percent of the total activity in the product. So, for example, if we consider a ^{201}Tl product that is assumed to be 97% ^{201}Tl, 2% ^{200}Tl, and 1% ^{202}Tl, we need to add the dose contributions from the primary product and the two contaminants. The easiest way to proceed is to calculate the dose per unit activity of each *pure product*, that is, the mGy per MBq of administered ^{201}Tl chloride, the mGy per MBq of administered ^{200}Tl chloride, and the mGy per MBq of administered ^{202}Tl chloride. Then we add 0.97 times the dose from the ^{201}Tl chloride to 0.02 times the dose from the ^{200}Tl chloride and 0.01 times the dose from the ^{202}Tl chloride. To develop the dose estimates for the contaminants, one usually employs the metabolic model for the primary nuclide and just changes the physical half-life to that for the contaminant, then recalculates the numbers of disintegrations for each of the contaminants. In the case in which a contaminant is not an isotope of the original nuclide, a separate dose estimate must be developed for this pharmaceutical from some available metabolic model. For example, ^{203}Hg also has been detected in ^{201}Tl chloride products. A possible metabolic model is that presented in ICRP Publication 30[17] for elemental mercury; this would assume that the mercury, when administered, does not stay with the ^{201}Tl chloride but moves freely in the body as elemental mercury. One might also assume that it stays with the ^{201}Tl chloride and apply the biokinetics for thallous chloride with the ^{203}Hg, as was done with the ^{200}Tl and ^{202}Tl.

For the case described, we could develop the dose estimates for pure ^{201}Tl chloride, ^{200}Tl chloride, and ^{202}Tl chloride as shown in Table 5.19.

Table 5.19. Radiation dose estimates from ^{201}Tl and associated contaminants.

	Dose (mGy/MBq)		
	^{201}Tl	^{200}Tl	^{202}Tl
Adrenals	1.72E-02	5.28E-03	3.02E-02
Brain	2.99E-02	8.33E-03	7.36E-02
Breasts	1.72E-02	5.28E-03	3.02E-02
Gallbladder wall	1.72E-02	5.28E-03	3.02E-02
LLI wall	2.24E-01	7.69E-02	3.10E-01
Small intestine	3.05E-01	1.03E-01	4.13E-01
Stomach	1.05E-01	3.47E-02	1.47E-01
ULI wall	1.86E-01	6.26E-02	2.52E-01
Heart wall	1.83E-01	6.32E-02	2.42E-01
Kidneys	3.07E-01	1.03E-01	4.36E-01
Liver	4.31E-02	1.45E-02	6.09E-02
Lungs	1.72E-02	5.28E-03	3.02E-02
Muscle	1.72E-02	5.28E-03	3.02E-02
Ovaries	1.72E-02	5.28E-03	3.02E-02
Pancreas	1.72E-02	5.28E-03	3.02E-02
Red marrow	1.72E-02	5.28E-03	3.02E-02
Bone surfaces	1.72E-02	5.28E-03	3.02E-02
Skin	1.72E-02	5.28E-03	3.02E-02
Spleen	9.98E-02	3.22E-02	1.75E-01
Testes	1.57E-01	5.15E-02	4.55E-01
Thymus	1.72E-02	5.28E-03	3.02E-02
Thyroid	4.74E-01	1.55E-01	7.05E-01
Urinary bladder wall	1.36E-02	4.25E-03	2.25E-02
Uterus	1.72E-02	5.28E-03	3.02E-02
Total body	2.42E-02	7.65E-03	4.02E-02

Abbreviations: LLI, lower large intestine; ULI, upper large intestine.

Taking 0.97 times the value in the first column, 0.02 times the value in the second column, and 0.01 times the value in the third column, we obtain the estimates for the product with contaminants, as shown in Table 5.20.

TABLE 5.20. Total radiation dose estimates from ^{201}Tl and all contaminants.

	Dose (mGy/MBq)
Adrenals	1.71E-02
Brain	2.99E-02
Breasts	1.71E-02
Gallbladder wall	1.71E-02
LLI wall	2.22E-01
Small intestine	3.02E-01
Stomach	1.04E-01
ULI wall	1.84E-01
Heart wall	1.81E-01
Kidneys	3.04E-01
Liver	4.27E-02
Lungs	1.71E-02
Muscle	1.71E-02
Ovaries	1.71E-02
Pancreas	1.71E-02
Red marrow	1.71E-02
Bone surfaces	1.71E-02
Skin	1.71E-02
Spleen	9.92E-02
Testes	1.58E-01
Thymus	1.71E-02
Thyroid	4.70E-01
Urinary bladder wall	1.35E-02
Uterus	1.71E-02
Total body	2.40E-02

Abbreviations: LLI, lower large intestine; ULI, upper large intestine.

Development and Use of Radiopharmaceuticals for Therapy

Dosimetry for therapeutic agents is not fundamentally different than for diagnostic agents. It is possible that one may wish to calculate dose to a tumor volume, in addition to the doses calculated to the usual organs and tissues of the body. Tumor ROIs may be drawn just as any organ ROI is drawn, with conjugate view imaging methods applied to estimate the activity in the tumor mass as a function of time and the time-integral of activity. Once the number of disintegrations is known, tumor self-dose may be obtained using absorbed fractions and dose factors from a number of publications that have treated energy absorption in unit-density spheres or ellipsoids of different sizes.[18,19,20,21] The OLINDA/EXM code[1] uses the absorbed fractions of Ref. 13 and provides tumor self-dose factors and doses, given entry of the number of disintegrations assumed to occur within the unit-density sphere. Table 5.21 shows values for ^{131}I, given a number of disintegrations of 1 MBq-h/MBq administered.

It is tempting to interpolate linearly in this table to obtain intermediate values, but this practice is generally discouraged, as the values often change substantially between table entries. A safer practice is to fit the values to a polynomial or multiple exponential functions and calculate the intermediate values from the function. For example, the data given in Table 5.21 (mGy/MBq) from 1 g to 100 g were fit to a three-component exponential function (Fig. 5.10).

The function was

$$D(m) = 188\,e^{-0.088m} + 28.7\,e^{-0.128m} + 4.74\,e^{-0.01355m}$$

For a mass of 15 g, this function gives a dose value of 8.09 mGy/MBq. Linear interpolation yields:

$$11.7 - \frac{15-10}{20-10}(11.7-5.94) = 8.82\frac{\text{mGy}}{\text{MBq}}$$

TABLE 5.21. Radiation dose estimates for ^{131}I in unit density spheres.

Self-dose from ^{131}I in unit-density spheres		
	Dose	
Sphere mass (g)	(mGy/MBq)	rad/mCi
0.01	9.68E+03	3.58E+04
0.1	1.04E+03	3.86E+03
0.5	2.14E+02	7.91E+02
1	1.11E+02	4.12E+02
2	5.62E+01	2.08E+02
4	2.85E+01	1.06E+02
6	1.92E+01	7.11E+01
8	1.45E+01	5.35E+01
10	1.17E+01	4.32E+01
20	5.94E+00	2.20E+01
40	3.03E+00	1.12E+01
60	2.05E+00	7.58E+00
80	1.56E+00	5.75E+00
100	1.26E+00	4.65E+00
300	4.43E-01	1.64E+00
400	3.39E-01	1.25E+00
500	2.75E-01	1.02E+00
600	2.31E-01	8.56E-01
1000	1.44E-01	5.33E-01
2000	7.63E-02	2.82E-01
3000	5.29E-02	1.96E-01
4000	4.10E-02	1.52E-01
5000	3.34E-02	1.24E-01
6000	2.84E-02	1.05E-01

Also, it is common in cancer patients to encounter organs that are notably different than the standard models, due to the disease and/or complications thereof. As noted in Chapter 3, the reported dose using a standard model may be adjusted for mass, scaling the electron and photon dose contributions separately. For electrons, the scaling is

$$DF_2 = DF_1 \frac{m_1}{m_2}$$

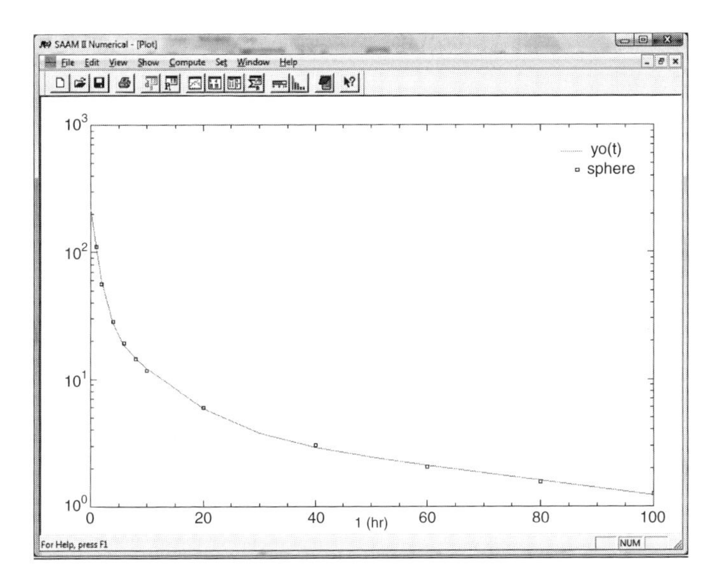

FIGURE 5.10. Fit to tumor dose factor data. (Created using SAAM II software: Simulation, Analysis, and Modeling Software for Kinetic Analysis. Software Copyright © 1992–2007 University of Washington, Seattle, WA. All rights reserved. For more information, see http://depts.washington.edu/saam2/.)

Here, DF_1 and DF_2 are the dose factors appropriate for use with organ masses m_1 and m_2. For photons, the scaling is

$$\phi_2 = \phi_1 \left(\frac{m_2}{m_1}\right)^{1/3} \qquad \Phi_2 = \Phi_1 \left(\frac{m_1}{m_2}\right)^{2/3}$$

To perform this calculation, you must isolate the Δ values for penetrating and nonpenetrating emissions, multiply them by these new absorbed fractions, and then recalculate the total dose by adding the two components together. Fortunately, the OLINDA/EXM code[1] performs this calculation automatically for the user, given entry of the new organ mass of interest.

The Pregnant Patient

Dose calculations for the pregnant or potentially pregnant subject are rather frequently encountered. In Chapter 3, the tables of dose estimates at various stages of pregnancy developed by Russell et al.[22] were presented. All published fetal dose estimates contain considerable uncertainties, and these are no exception. One should estimate fetal dose carefully, understand the model used to develop the numerical estimates of dose, and use discretion and conservatism in interpreting the results. During the first, say, 3 to 6 weeks, the dose to the nongravid uterus (the "early pregnancy" dose) is probably a good estimate of the dose to the fetus. In other cases involving later pregnancy, one can look at the dose on either side of the actual time of gestation and just use what is believed to be the most appropriate, most conservative, and so forth, estimate. Of particular concern are the issues of:

- Potential placental crossover of the radiopharmaceutical: Some of the pharmaceuticals treated in the Russell et al.[22] tables included some consideration of placental crossover in the dose estimates (usually based on results of animal studies), whereas others did not treat this issue, as no information was available. Whether or not placental crossover occurs and, if so, in what concentrations it may occur, can profoundly affect the dose estimates to the fetus.
- Dose to the fetal thyroid for cases involving radioiodines: After about the 10th to 13th week of pregnancy, iodine begins to be concentrated in the fetal thyroid, which is a very small organ that can easily receive a very high dose from radiopharmaceutical administrations.[23] One is usually interested in the average dose to the fetus for risk evaluation; in the case of radioiodines involved in studies with women at this or later stages of pregnancy, it is important to also consider the fetal thyroid dose.

Here is an example. A woman receives 750 MBq of ^{99m}Tc-MDP for a bone scan. Later, it is found out that she was about 1 week pregnant at the time of the scan. Here, the most appropriate fetal dose is $750\,\text{MBq} \times 0.0061\,\text{mGy/MBq} = 4.6\,\text{mGy} \approx 5\,\text{mGy}$.

Another example follows. A woman receives 200 MBq of ^{99m}Tc-MAA for a lung scan at about 7 months' gestation. This is not unusual: Some women have a tendency to form blood clots in later pregnancy, and lung scans are often performed on patients known to be pregnant. The dose estimate at 6 months is $200\,\text{MBq} \times 0.005\,\text{mGy/MBq} = 1\,\text{mGy}$, and the estimate at 9 months is $200\,\text{MBq} \times 0.004\,\text{mGy/MBq} = 0.8\,\text{mGy}$. Given the uncertainty and the time, the dose estimate should be cited as 1 mGy.

As shown in the following example, the estimates of the numbers of disintegrations in source organs in Russel et al.[22] may be used in cases in which the user happens to know something about changes in such values in individual subjects. In this case, a more careful dose calculation may be made using available DF values.

A woman in early pregnancy is administered 750 MBq ^{99m}Tc-DTPA, but due to a kidney being blocked, the number of disintegrations in kidney is thought to be much higher (around 2.5 MBq-h/MBq) than in the standard model (0.92 MBq-h/MBq). No estimate is made of the bladder or remainder of the body residence time. The standard dose estimate, as given in the tables, is (the DF values were taken from adult female model in the OLINDA/EXM 1.0 software[1]):

$$\begin{aligned}
D_{\text{fetus}} = A_0[&N_{\text{kidneys}}\text{DF}(\text{uterus} \leftarrow \text{kidneys}) + \\
&N_{\text{bladder}}\text{DF}(\text{uterus} \leftarrow \text{bladder}) + \\
&N_{\text{remainder}}\text{DF}(\text{uterus} \leftarrow \text{remainder})]
\end{aligned}$$

$$\begin{aligned}
D_{\text{fetus}} = 750\,\text{MBq}[&0.092\,\text{MBq-h/MBq} \times 3600\,\text{s/h} \times 8.44 \times 10^{-8} \\
&\text{mGy/MBq-s} + 1.84\,\text{MBq-h/MBq} \times 3600\,\text{s/h} \times \\
&1.48 \times 10^{-6}\text{mGy/MBq-s} + 2.84\text{MBq-h/MBq} \times \\
&3600\,\text{s/h} \times 2.14 \times 10^{-7}\text{mGy/MBq-s}]
\end{aligned}$$

$$D_{\text{fetus}} = 9.0\,\text{mGy}$$

The new dose, modified for this subject, is

$$
\begin{aligned}
D_{\text{fetus}} = & A_0[N_{\text{kidneys}}\text{DF}(\text{uterus} \leftarrow \text{kidneys})+ \\
& N_{bladder}\text{DF}(\text{uterus} \leftarrow \text{bladder})+ \\
& N_{\text{remainder}}\text{DF}(\text{uterus} \leftarrow \text{remainder})] \\
D_{\text{fetus}} = & 750\,\text{MBq}[2.5\,\text{MBq-h/MBq} \times 3600\,\text{s/h} \times 8.44 \times 10^{-8} \\
& \text{mGy/MBq-s} + 1.84\,\text{MBq-h/MBq} \times 3600\,\text{s/h} \times \\
& 1.48 \times 10^{-6}\,\text{mGy/MBq-s} + 2.84\,\text{MBq-h/MBq} \times \\
& 3600\,\text{s/h} \times 2.14 \times 10^{-7}\,\text{mGy/MBq-s}] \\
& D_{\text{fetus}} = 9.6\,\text{mGy}
\end{aligned}
$$

This is a very small change in the dose estimate, which is not surprising, given the small magnitude of the kidney to uterus S value. Let's consider the situation in which the same woman was encouraged to void her bladder more frequently than in the standard dose estimate for ^{99m}Tc-DTPA, and her bladder residence time decreased to 0.9 hour:

$$
\begin{aligned}
D_{\text{fetus}} = & A_0[N_{\text{kidneys}}\text{DF}(\text{uterus} \leftarrow \text{kidneys}) \\
& + N_{\text{bladder DF}}(\text{uterus} \leftarrow \text{bladder}) \\
& + N_{\text{remainder}}\text{DF}(\text{uterus} \leftarrow \text{remainder})] \\
D_{\text{fetus}} = & 750\,\text{MBq}[2.5\,\text{MBq-h/MBq} \times 3600\,\text{s/h} \times 8.44 \times 10^{-8} \\
& \text{mGy/MBq-s} + 0.90\,\text{MBq-h/MBq} \times 3600\,\text{s/h} \\
& \times 1.48 \times 10^{-6}\,\text{mGy/MBq-s} + 2.84\,\text{MBq-h/MBq} \\
& \times 3600\,\text{s/h} \times 2.14 \times 10^{-7}\,\text{mGy/MBq-s}] \\
& D_{\text{fetus}} = 5.8\text{ mGy}
\end{aligned}
$$

References

1. Stabin MG, Sparks RB, Crowe E. OLINDA/EXM: the second-generation personal computer software for internal dose assessment in nuclear medicine. J Nucl Med 46:1023–1027, 2005.

2. Cloutier R, Watson E, Rohrer R, Smith E. Calculating the radiation dose to an organ. J Nucl Med 14:53–55, 1973.
3. International Commission on Radiological Protection. Limits for Intakes of Radionuclides by Workers. ICRP Publication 30. Pergamon Press, New York, 1979.
4. International Commission on Radiological Protection. No. 70. Basic Anatomical and Physiological Data for Use in Radiological Protection: the Skeleton. Pergamon Press, Elmsford, NY, 1997, p. 23.
5. Cloutier R, Smith S, Watson E, Snyder W, Warner G. Dose to the fetus from radionuclides in the bladder. Health Phys 25:147–161, 1973.
6. Stabin MG, da Luz CQPL. New decay data for internal and external dose assessment. Health Phys 83:471–475, 2002.
7. Abramoff MD, Magelhaes PJ, Ram SJ. Image processing with ImageJ. Biophotonics Int 11:36–42, 2004.
8. Sjogreen K, Ljungberg M, Strand S-K. An activity quantification method based on registration of CT and whole-body scintillation camera images, with application to 131I. J Nucl Med 43:972–982, 2002.
9. Sgouros G. Bone marrow dosimetry for radioimmunotherapy: theoretical consideration. J Nucl Med 93; 34:689–694, 1993.
10. Cremonesi M, Ferrari M, Bodei L, Tosi G, Paganelli G. Dosimetry in peptide radionuclide receptor therapy: a review. J Nucl Med 47:1467–1475, 2006.
11. Siegel JA, Lee RE, Pawlyk DA et al. Sacral scintigraphy for bone marrow dosimetry in radioimmunotherapy. Int. J Rad Appl Instrum B16:553–559, 1989.
12. Cristy M. Active bone marrow distribution as a function of age in humans. Phys Med Biol 26:389–400, 1981.
13. International Commission on Radiological Protection. ICRP Publication 89. Basic Anatomical and Physiological Data for Use in Radiological Protection: Reference Values. Pergamon Press, Elmsford, NY, 2003.
14. Foster D, Barrett P. Developing and testing integrated multicompartment models to describe a single-input multiple-output study using the SAAM II software system. In: Proceedings of the Sixth International Radiopharmaceutical Dosimetry Symposium. Watson EE, Schlafke-Stelson AT, eds, Oak Ridge Institute for Science and Education, Oak Ridge, TN, 1999, pp. 577–599.
15. Watson EE, Stabin MG, Davis JL, Eckerman KF. A model of the peritoneal cavity for use in internal dosimetry. J Nucl Med 30:2002–2011, 1989.

16. Stabin MG, Siegel JA. Physical models and dose factors for use in internal dose assessment. Health Phys 85:294–310, 2003.
17. International Commission on Radiological Protection. Limits for Intakes of Radionuclides by Workers. ICRP Publication 30. Pergamon Press, New York, 1979.
18. Brownell G, Ellett W, Reddy R. MIRD Pamphlet No. 3: absorbed fractions for photon dosimetry. J Nucl Med (Suppl 1):27, 1968.
19. Ellett W, Humes R. MIRD Pamphlet No. 8: absorbed fractions for small volumes containing photon-emitting radioactivity. J Nucl Med (Suppl 6):7, 1972.
20. Siegel JA, Stabin MG. Absorbed fractions for electrons and beta particles in spheres of various sizes. J Nucl Med 35: 152–156, 1994.
21. Stabin MG, Konijnenberg M. Re-evaluation of absorbed fractions for photons and electrons in small spheres. J Nucl Med 41:149–160, 2000.
22. Russell JR, Stabin MG, Sparks RB, Watson EE. Radiation absorbed dose to the embryo/fetus from radiopharmaceuticals. Health Phys 73:756–769, 1997.
23. Watson EE. Radiation Absorbed Dose to the Human Fetal Thyroid. In: Fifth International Radiopharmaceutical Dosimetry Symposium. Stelson A, Stabin M, Sparks R, eds, Oak Ridge Associated Universities, Oak Ridge, TN, 1992, pp. 179–187.

6 Biological Effects of Radiation*

Basic Principles

Very soon after the discovery of radiation and radioactivity, it became evident that exposure to radiation could induce short-term and long-term negative effects in human tissue. The first possible adverse effects of X-rays were observed by Thomas Edison, William J. Morton, and Nikola Tesla. These investigators independently reported eye irritations from experimentation with X-rays and fluorescent substances. The effects were first attributed to eye strain or, possibly, ultraviolet radiation from the long-term direct observance of fluorescence. Elihu Thomson (an American physicist) deliberately exposed the little finger of his left hand to an X-ray tube for several days, for about half an hour per day. This resulted in pain, swelling, stiffness, erythema, and blistering in the finger, which was clearly and immediately related to the radiation exposure. William Herbert Rollins (a Boston dentist) showed that X-rays could kill guinea pigs and result in the death of offspring when guinea pigs were irradiated while pregnant. In 1898, Henri Becquerel received a skin burn from a radium source given to him by the Curies that he kept in his vest pocket for some time. He carried the source

*Portions of this chapter are reproduced and adapted, with permission, from Stabin, M. Radiation Protection and Dosimetry. Chapter 6: Biological effects of radiation. Springer, New York, 2007.

with him on his travels, to use it in demonstrations during his lectures. He declared, "I love this radium but I have a grudge against it!" The first death in an X-ray pioneer attributed to cumulative overexposure was that of C.M. Dally in 1904. It was later observed that radiologists and other physicians who used X-rays in their practices before health physics practices were common had a significantly higher rate of leukemia than their colleagues.

A particularly tragic episode in the history of the use of radiation and in the history of industrialism was the acute and chronic damage done to the radium dial painters.[1] Radium was used in luminous paints in the early 1900s. In factories where luminous dial watches were made, workers (mainly women) would sharpen the tips of their paint brushes with their lips and ingested large amounts of radium. They had increased amounts of bone cancer (carcinomas in the paranasal sinuses or the mastoid structures, which are very rare, and were thus clearly associated with their exposures, as well as cancers in other sites) and even spontaneous fractures in their jaws and spines from cumulative radiation injury. Others died of anemia and other causes.

Mechanisms of Radiation Damage to Biological Systems

Radiation interactions with aqueous systems can be described as occurring in four principal stages:

1. Physical
2. Prechemical
3. Early chemical
4. Late chemical

In the *physical* stage of water radiolysis, a primary charged particle interacts through elastic and inelastic collisions. Inelastic collisions result in the ionization and excitation of water molecules, leaving behind ionized (H_2O^+) and excited (H_2O^*) molecules and unbound subexcitation electrons (e^-_{sub}). A subexcitation electron is one whose energy is not high enough to produce further electronic transitions. By contrast, some electrons produced in the interaction of the primary charged particle with the water molecules may have sufficient energy themselves to produce additional electronic transitions. These electrons may produce secondary track structures (delta rays), beyond that produced by the primary particle. All charged particles can interact with electrons in the water both individually and collectively in the condensed, liquid phase. The initial passage of the particle, with the production of ionized and excited water molecules and subexcitation electrons in the local track region (within a few hundred angstroms), occurs within about 10^{-15} s. From this time until about 10^{-12} s, in the *prechemical* phase, some initial reactions and rearrangements of these species occur. If a water molecule is ionized, this results in the creation of an ionized water molecule and a free electron. The free electron rapidly attracts other water molecules, as the slightly polar molecule has a positive and negative pole, and the positive pole is attracted to the electron. A group of water molecules thus clusters around the electron, and it is known as a *hydrated electron* and is designated as e_{aq}^-. The water molecule dissociates immediately:

$$H_2O \rightarrow H_2O^+ + e_{aq}^- \rightarrow H^+ + OH\cdot + e_{aq}^-$$

In an excitation event, an electron in the molecule is raised to a higher energy level. This electron may simply return to its original state, or the molecule may break up into an H and an OH radical. (A radical is a species that has an unpaired electron in one of its orbitals: the species is not necessarily charged but is highly reactive.)

$$H_2O \rightarrow H\cdot + OH\cdot$$

The free radical species and the hydrated electron undergo dozens of other reactions with each other and other molecules in the system. Reactions with other commonly encountered molecules in aqueous systems are shown in Table 6.1. Reactions with other molecules have been studied and modeled by various investigators as well.[2–5]

TABLE 6.1. Comparison of reaction rate coefficients and reaction radii for several reactions of importance to radiation biology.

Reaction	k ($10^{10}\,M^{-1}\,s^{-1}$)	R (nm)
$H\cdot + OH\cdot \rightarrow H_2O$	2.0	0.43
$e_{aq}^- + OH\cdot \rightarrow OH^-$	3.0	0.72
$e_{aq}^- + H\cdot + H_2O \rightarrow H_2 + OH^-$	2.5	0.45
$e_{aq}^- + H_3O^+ \rightarrow H\cdot + H_2O$	2.2	0.39
$H\cdot + H\cdot \rightarrow H_2$	1.0	0.23
$OH\cdot + OH\cdot \rightarrow H_2O_2$	0.55	0.55
$2e_{aq}^- + 2H_2O \rightarrow H_2 + 2OH^-$	0.5	0.18
$H_3O^+ + OH^- \rightarrow 2H_2O$	14.3	1.58
$e_{aq}^- + H_2O_2 \rightarrow OH^- + OH\cdot$	1.2	0.57
$OH\cdot + OH^- \rightarrow H_2O + O^-$	1.2	0.36

Note: k is the reaction rate constant and R is the *reaction radius* for the specified reaction. Use of these concepts is explained in the physical chemistry literature.

The *early chemical* phase, extending from $\sim 10^{-12}$ s to $\sim 10^{-6}$ s, is the time period within which the species can diffuse and react with each other and with other molecules in solution. By about 10^{-6} s, most of the original track structure is lost, and any remaining reactive species are so widely separated that further reactions between individual species are unlikely.[5] From 10^{-6} s onward, referred to as the *late chemical* stage, calculation of further product yields can be made by using differential rate-equation systems that assume uniform distribution of the solutes and reactions governed by reaction-rate coefficients (Fig. 6.1).

Observed Biological Effects in Humans

There are two broad categories of radiation-related effects in humans: *stochastic* and *nonstochastic*. There are three important characteristics that distinguish them.

Nonstochastic effects (now officially called *deterministic effects*, previously also called *acute effects*) are effects that are generally observed soon after exposure to radiation. As they are "nonstochastic" in nature, they will always be observed (if the dose threshold is exceeded), and there is generally no doubt that they were caused by the radiation exposure.

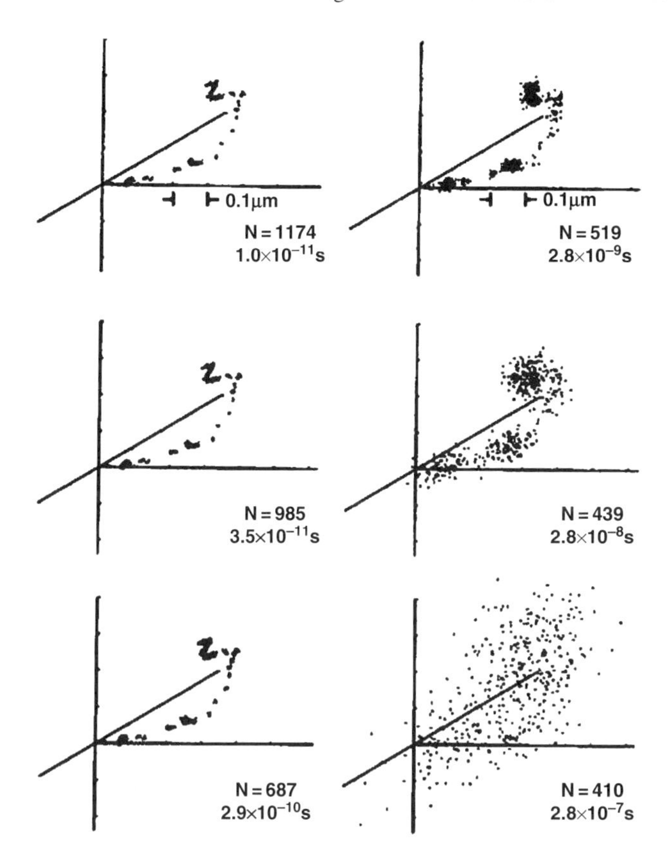

FIGURE 6.1. Chemical development of an electron track over the first $\sim 10^{-6}$ seconds after passage of the electron. (Reproduced with permission from Turner J. Atoms, Radiation, and Radiation Protection. Pergamon Press, New York, 1986; and with permission of the Oak Ridge National Laboratory, managed by UT-Battelle, LLC, for the U.S. Department of Energy.)

The major identifying characteristics of nonstochastic effects are

1. There is a *threshold* of dose below which the effects will not be observed.
2. Above this threshold, the *magnitude* of the effect increases with dose.
3. The effect is *clearly associated* with the radiation exposure.

Examples of these effects are

- Erythema (reddening of the skin)
- Epilation (loss of hair)
- Depression of bone marrow cell division (observed in counts of formed elements in peripheral blood)
- NVD (nausea, vomiting, diarrhea), often observed in victims after an acute exposure to radiation
- Central nervous system damage
- Damage to the unborn child [physical deformities, microcephaly (small head size at birth), mental retardation]

When discussing nonstochastic effects, it is important to note that some organs are more radiosensitive than others. The so-called law of Bergonie and Tribondeau[6] states that cells tend to be radiosensitive if they have three properties:

- Cells have a high division rate.
- Cells have a long dividing future.
- Cells are of an unspecialized type.

A concise way of stating the law might be to say that the radiosensitivity of a cell type is proportional to its rate of division and inversely proportional to its degree of specialization. So, rapidly dividing and unspecialized cells, as a rule, are the most radiosensitive. Two important examples are cells in the red marrow and in the developing embryo/fetus; in the case of marrow, a number of *progenitor* cells that, through many generations of cell division, produce a variety of different functional cells that are very specialized (e.g., red blood cells, lymphocytes, leukocytes, platelets) (Fig. 6.2). Some of these functional cells do not divide at all and are thus themselves quite radioresistant. However, if the marrow receives a high dose of radiation, damage to these progenitor cells is very important to the health of the organism. As we will see shortly, if these cells are affected, in a brief period this will be manifested in a measurable decrease in the number of formed elements in the peripheral blood. If the damage is severe enough, the person may not survive. If not, the progenitor cells will eventually repopulate and begin to replenish the numbers of the formed elements, and subsequent blood samples will show this recovery process.

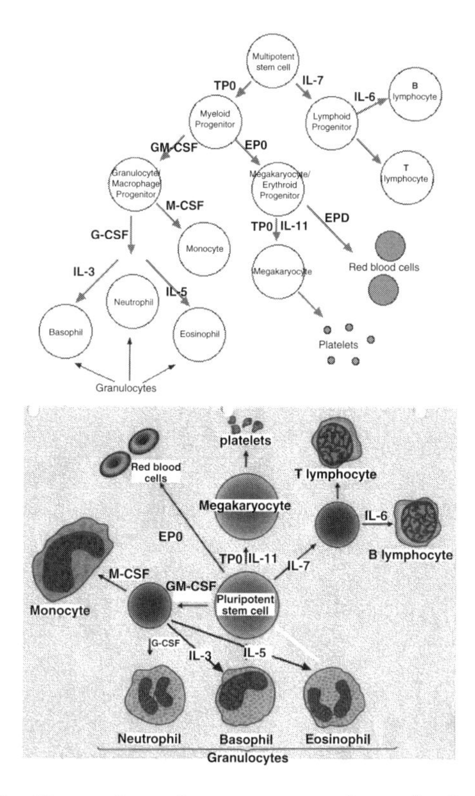

FIGURE 6.2. Generations of marrow progenitor cells. (Reproduced with permission from Kimball's Biology Pages: http://users.rcn.com/jkimball.ma.ultranet/BiologyPages/B/Blood.html# formation.)

In the fetus, organs and systems develop at different rates. At the moment of conception, of course, we have one completely undifferentiated cell that becomes two cells after one division, then four, then eight, and so on. As the rapid cell division proceeds, groups of cells "receive their assignments" and differentiate to form organs and organ systems, still with a very rapid rate of cell division. At some point, individual organs become well defined and formed, and cell division slows as the fetus simply adds mass. But while differentiation and early rapid cell division is occurring, these cells are quite radiosensitive, and a high dose to the fetus may cause fetal death or damage to individual fetal structures. This is discussed later in this chapter. On the other hand, in an adult, cells of the central nervous system (CNS; brain tissue, spinal cord, etc.) are very highly specialized and have very low or no rate of division. The CNS is thus particularly radioresistant. One important nonstochastic effect is death. This results from damage to the bone marrow (first), then to the gastrointestinal tract, then to the nervous system.

Stochastic effects are effects that are, as the name implies, probabilistic. They may or may not occur in any given exposed individual. These effects generally manifest many years, even decades, after the radiation exposure (and were once called *late effects*). Their major characteristics, in direct contrast with those for nonstochastic effects, are (1) A *threshold* may not be observed. (2) The *probability* of the effect increases with dose. (3) You *cannot definitively associate* the effect with the radiation exposure. Examples of these effects include cancer induction and genetic effects (offspring of irradiated individuals).

Cancer

The fact that ionizing radiation causes cancer is well established. Exposures of a number of populations, in addition to animal studies, have established clear causative links between radiation exposure and expression of a number of types of cancer. The quantitative relationship is sometimes fairly well established, in other cases less well. At high enough doses, the rate of production clearly increases with increasing dose (i.e., probability increases with dose). A radiation-induced cancer is indistinguishable from a "spontaneous" cancer; the causal link is established from the number of cancers induced in an exposed population in relation to that expected in that population otherwise. In the populations that have been studied to establish relationships between dose and risk, in most cases, the radiation doses themselves and the rates of cancer are subject to considerable uncertainty. The most important population studied to date to determine these trends is the population of survivors of the atomic bomb attacks on Japan at the end of World War II. Several hundred thousand people died either instantly or within the first year after the attacks from physical injuries and radiation sickness. The surviving population has been extensively studied over the years after the attacks. The most important single institution in this follow-up effort is the Radiation Effects Research Foundation (RERF), with locations in both Hiroshima and Nagasaki.[7] The RERF (formerly the Atomic Bomb Casualty Commission) was founded in April 1975 and is a private, nonprofit Japanese foundation. Funding is provided by the governments of Japan, through the Ministry of Health, Labor, and Welfare, and by the Department of Energy in the United States. Some 36,500 survivors who were exposed beyond 2.5 km have been followed medically continuously since the blasts. Of this population, about 4900 cancer deaths have been identified, including about 180 leukemia deaths and 4700 deaths from cancers other than leukemia. Of these, only about 89 and 340 deaths, respectively, appear to be attributable to radiation. There are also a number of populations of individuals exposed to various medical studies using high levels of radiation (principally from the early 1900s, before radiation's dangers were fully appreciated). For different types of cancer, if we plot the cancer rate against the dose received, the data will show an upward trend. The question that none of the data sets clearly answers is that of the shape of the curve at low doses and dose rates. A controversy continues to rage over whether the relationship between dose and the absolute or relative number of induced cancers should be extrapolated to zero dose (i.e., all exposure to radiation, no matter how small, is associated with some risk of cancer) or if there is a *threshold* (Fig. 6.3). There is evidence to support both views. In fact, there is some evidence to support the controversial theory of *hormesis*: that exposure to low levels of radiation is associated with *less* cancer induction than in systems deprived of all radiation exposure. The possible mechanism here is that exposure to radiation stimulates cellular repair mechanisms (the mechanism of *adaptive response*). In individual

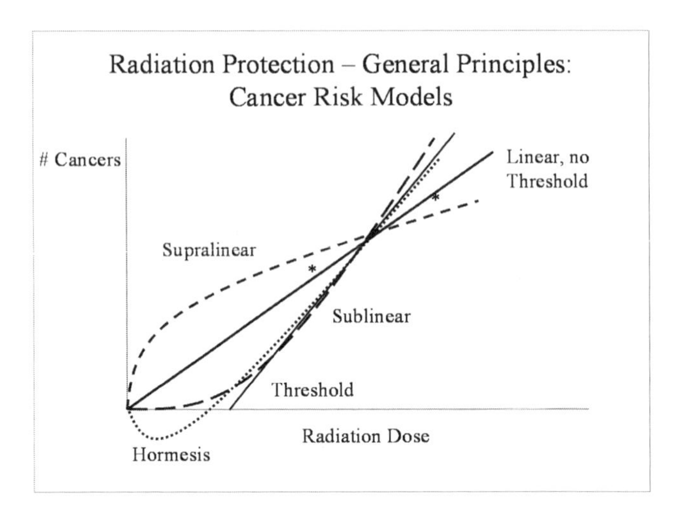

FIGURE 6.3. General principles of radiation risk models.

experiments, hormetic and adaptive response mechanisms have been demonstrated.[8]

Cell Survival Studies

Thus far, we have looked at gross effects on the organism or individual tissues of the organism after exposure to radiation. Much information on the biological effects of radiation has been obtained for many years through the use of direct experiments on cell cultures. It is, of course, far easier to control the experiment and the variables involved when the radiation source can be carefully modulated, the system under study can be simple and uniform, and the results can be evaluated over most any period of time desired (days to weeks, or even over microseconds, such as in the study of free radical formation and reaction[9]). After exposure of a group of cells to radiation, the most common concept to study is that of cell survival. Typically, the natural logarithm of the surviving fraction of irradiated cells is plotted against the dose received (Fig. 6.4).

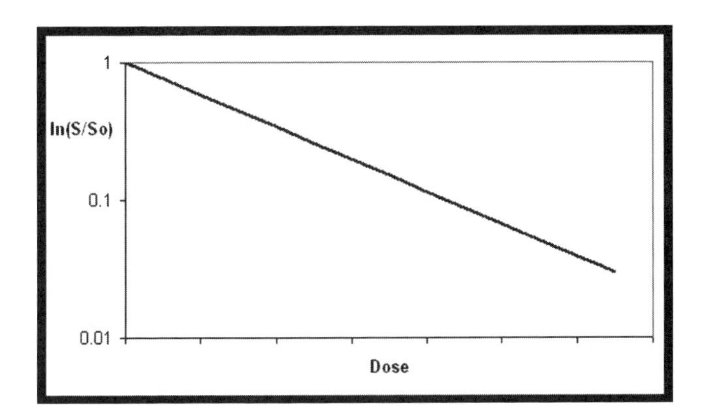

FIGURE 6.4. Typical cell survival curve after exposure to high LET radiation.

The simplest survival curve is a single exponential:

$$S = S_0\, e^{-D/D_0}$$

Here, S is the surviving fraction, S_0 is the original number of cells irradiated, D is the dose received, and D_0 is the negative reciprocal of the slope of the curve and is called the mean lethal dose. When cells receive dose D_0, the surviving fraction is 0.37, which is $1/e$. This dose may also be referred to as the D_{37} dose, just as we define the LD_{50}, the lethal dose of radiation that will kill half of a population. Generally speaking, particles with a high linear energy transfer (LET) will show this form of a survival curve, whereas those of low LET will have a more complicated curve, of the form

$$S = S_0\left[1 - (1 - e^{-D/D_0})^n.\right]$$

Here, n is the assumed number of targets that need to be hit in order to inactivate a cell. If $n = 1$, the equation reduces to the more simple form shown above. The usual curve, however, has a "shoulder," indicating that a certain amount of dose must be received before any significant effect on cell survival is seen. At higher doses, the curve attains the usual linear shape with slope $-1/D_0$. If the linear portion is extrapolated back to zero dose, it will intercept the y axis at the *extrapolation number*, n, which is numerically equal to the number of targets assumed to be relevant to the cells' survival (Fig. 6.5).

Several factors affect the shape of the dose-response function other than the LET of the radiation, including:

- *Dose rate*: The LD_{50} of a population of cells will clearly increase as the dose rate at which a fixed dose D is delivered is decreased. Cells have a considerable capacity to repair radiation damage, and, if time is allowed for repair, more radiation can be tolerated.
- *Dose fractionation*: If cells are given a cumulative dose D, but instead of being delivered all at once, it is delivered in N fractions of D/N each, the cell survival curve will show a *series of shoulders* linked together, because cellular repair is again ongoing between fractions. This is a strategy used in radiation therapy procedures to allow healthy tissues time for repair while still delivering an ultimately lethal dose to the tumor tissues.
- *Presence of oxygen*: Dissolved oxygen in tissue causes the tissue to be sensitive to radiation. Hypoxic cells have been shown to be considerably more radioresistant. The effect of oxygen is sometimes expressed as the *oxygen enhancement*

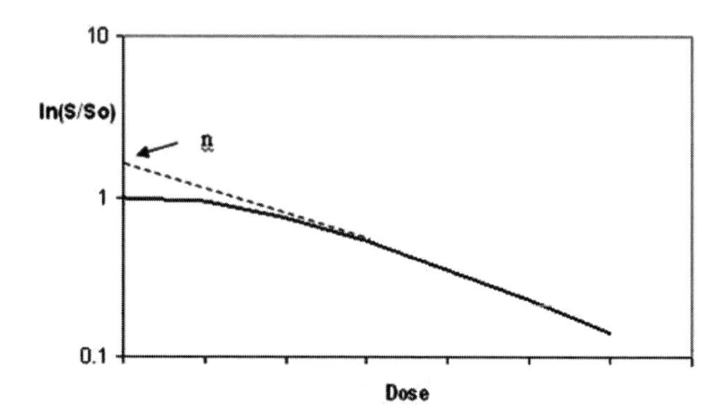

FIGURE 6.5. Typical cell survival curve after exposure to low LET radiation.

ratio (OER), which is the ratio of the slope of the straight portion of the cell survival curve with and without oxygen present.

Relative Biological Effectiveness

The exact form of a dose-response curve depends on the type of cells irradiated, the type and energy of radiation used, and the biological end point studied. Thus far, we have only mentioned cell inactivation or death as an end point, but other end points as well may be considered, such as a specific level of cell killing (37%, 10% survival, etc.), oncogenic transformation, or induction of chromosomal aberrations. Historically, the most studied form of radiation in cell studies were 250 kVp X-rays. When the effects of other radiations on the same cell population to produce the same end point were studied, it was quickly seen that all radiation does not produce the same effects at the same dose levels as this "reference" radiation. If a dose D' of a given radiation type produces the same biological end point in a given experiment as a dose D of our reference radiation, we can define a quantity called the relative biological effectiveness (RBE)[10] as:

$$\mathrm{RBE} = \frac{D}{D'}$$

So, for example, if a dose of 1 Gy of the reference radiation produces a particular cell survival level, but only 0.05 Gy of alpha radiation produces the same level of cell killing, we say that the RBE for alpha particles in this experiment is 20.

RBE is quite dependent on radiation LET. High LET radiations generally have high RBEs; you should note that 250 kVp X-rays are generally considered to be low LET radiation. The relationship of the two variables is not directly linear, but there is clearly a positively correlated relationship of RBE with LET, until very high LET values are reached, where "overkill" of cells causes the RBE not to increase as quickly.

You may have noted that, in the numerical example chosen above, the RBE for alpha particles is exactly equal to the currently recommended value of w_R, the radiation weighting factor used in radiation protection. This was quite intentional. Values of w_R are very closely tied to RBE values, however, they are *not* exactly equal. Generally, conservative values of RBE were used to set the values assigned for w_R values (also formerly called *quality factors*, you may recall). The important thing to remember about RBE values is that they are *highly dependent on the experimental conditions* (cell type, radiation type, radiation dose rate) and the biological end point defined for study. Radiation weighting factors, on the other hand, are *single values* to be applied to a type of radiation in all situations. Radiation weighting factors are *operational* quantities, used to solve a practical problem (how to best protect radiation workers from routine exposure to radiation), whereas RBEs are more *scientific* quantities relevant to the study of radiation biology.

Cells clearly have mechanisms for repairing DNA damage. If damage occurs to a single strand of DNA, it is particularly easy for the cells to repair this damage, as information from the complementary chain may be used to identify the base pairs needed to complete the damaged area. *Double strand breaks* are more difficult to repair, but cellular mechanisms do exist that can affect repair here also.

New Evidence—and Some Confusion!

On the other hand, very recent research has profoundly challenged conventional notions of the relationship between radiation dose and observed effects. In a report by the American Association of Physicists in Medicine (AAPM),[11] it is noted that various cellular and organ studies have shown that low-LET-type effects have been seen when Auger emitters are present only in the cytoplasm of cells, whereas when Auger emitters may be incorporated into the *DNA of cells*, the resulting survival curves are similar to those normally seen for high-LET alpha particles, as shown in Figure 6.4. Also, in some in vivo studies with radioprotectors, the intense local damage imparted to cells by Auger emissions has been shown to be able to be mitigated somewhat even though the high-LET-type effects are known to be present. The committee reached the rather unsettling conclusion that[11]:

> The absorbed dose from Auger emitters must be calculated at a level suited to the biological system employed. Hence, a number of target volumes are of interest:
>
> - individual strands or bases of the DNA molecule
> - supercoiled DNA
> - cell or cell nucleus
> - bulk tissue
>
> Choosing the target volume, however, is complex. The radiation properties of the radionuclide certainly play a role in this regard. Just as important is the distribution of the radioactivity within the cells, which in turn depends on the chemical nature of the radiocompound. Hence, *the appropriate target volume must be determined on a case-by-case basis*. [emphasis added.][†]

The group ultimately concluded that a value of 10 be used for the radiation weighting factor (w_R) for predicting therapeutic outcome if an Auger emitter is used that is thought to be covalently bound to the DNA of the cells treated. If the emitter, conversely, is localized in the nucleus, but not covalently bound to DNA, a weighting factor of 5 was recommended.

Brooks points out that absorbed dose is often used too liberally as a direct indicator of radiation risk.[12] Whereas the simple concept of energy absorbed per unit mass of tissue has good predictive value at some dose levels and if activity is uniformly distributed throughout an organ (in the case of internal emitters), it is clearly not a good predictor of biological response when activity is not uniformly distributed

[†]Excerpt reprinted with permission from Humm, J., et al. "Dosimetry of Auger-electron-emitting radionuclides: Report No. 3 of AAPM Nuclear Task Group No. 6a." Medical Physics 21, 2004; 1994.

and when energy deposited by high LET particles occurs in regions where it is difficult to distribute the energy over the appropriate target mass (as noted by the AAPM, earlier in this chapter[11]). Recent experimental evidence has shown that energy distribution alone cannot always predict the occurrence of cellular changes, but that in some conditions, cells with no direct energy deposition from radiation may demonstrate a response (the *bystander effect*). Brooks[12] notes that "The potential for bystander effects may impact risk from nonuniform distribution of dose or energy in tissues and raises some very interesting questions as to the validity of such calculations." Hall[13] notes that "The plethora of data now available concerning the bystander effect fall into two quite separate categories, and it is not certain that the two groups of experiments are addressing the same phenomenon." Those two categories are

1. *Medium transfer experiments*: In a number of independent studies,[13] irradiated cells appear to have secreted some molecule into the culture medium that was capable of killing cells when that medium was placed in contact with unirradiated cells. The effect produced by epithelial cell cultures is dependent on the cell number at the time of irradiation and can be observed as soon as 30 minutes postirradiation, and may still be effective if taken from the irradiated cells up to 60 hours after irradiation. This bystander effect can be induced by radiation doses as low as 0.25 mGy and does not appear to be significantly increased up to doses of 10 Gy. In addition to increased levels of cell death and reduced cloning efficiency, medium transfer experiments have shown an increase in neoplastic transformation as well as genomic instability in cells that have not themselves been irradiated.
2. *Microbeam irradiation experiments*: In these studies, also reproduced by various investigators, the use of accurately directed beams of radiation permits the exposure of some cells in a culture medium to the radiation, but not others, and effects in the unirradiated cells have been clearly seen. Hall[13] discusses one of the more striking experiments, in which human fibroblasts were irradiated with

> microbeams of alpha particles, with cells of one population lightly stained with cyto-orange, a cytoplasmic vital dye, whereas cells of another population were lightly stained blue with a nuclear vital dye. The two cell populations were mixed and allowed to attach to the culture dish, and the computer controlling the accelerator was programmed to irradiate only blue-stained cells with 10 alpha particles directed at the centroid of the nucleus. The cells were fixed and stained 48 hours later, at which time micronuclei and chromosome bridges were visible in a proportion of the nonhit (i.e., orange-stained) cells!

Other striking studies have involved the irradiation of the lung base in rats, with a marked increase in the frequency of micronuclei found in the shielded lung apex.[13] However, radiation of the lung apex did not result in an increase in the chromosome damage in the shielded lung base. This suggests that a factor was transferred from the exposed portion of the lung to the shielded part and that this transfer has direction from the base to the apex of the lung. In another experiment, exposure of the left lung resulted in a marked increase in micronuclei in the unexposed right lung. Experiments suggest that bystander effects are limited to the organ irradiated and have been demonstrated primarily in experiments with alpha particles. These results challenge the traditional notion of the relationship of dose and effects.

Another mechanism, called *genomic instability*, also suggests that the effects from radiation may be felt in cells other than those directly irradiated. Cells irradiated with radiation have been shown to not have observable radiation damage, but subsequent generations of these cells may show DNA damage. Morgan notes that, while genomic instability has been demonstrated in vitro and in vivo, some results are conflicting, and interpretation remains controversial.[14] Morgan states that "Because radiation risk estimates are also organ specific, it is reasonable to assume that any bystander effect induced in vivo is accounted for in models of organ risk evaluation. As a result, it is unlikely that the resurgence of interest in these non-targeted radiation effects will substantially alter risk estimates."[14]

In a fascinating in vivo experiment, Hishikawa et al.[15] subcutaneously injected nude mice with mixtures of unirradiated human adenocarcinoma LS174T cells and cells that had been *lethally* irradiated with ^{125}I Auger electrons emitted in the DNA of the cells. As expected, they saw inhibition of tumor growth compared with mice exposed to similar mixtures of completely unirradiated cells. However, what was surprising was that when they repeated the experiment with cells that had been lethally irradiated with ^{123}I, a "stimulatory bystander effect" was seen; that is, *more* tumor growth was actually seen in the mice injected with the mixture of ^{123}I irradiated and unirradiated cells! A complete explanation of this phenomenon has not yet been offered.

Thus, whereas evidence from studies that purport to show hormesis and adaptive response suggest that low levels of radiation may not be harmful and may even be beneficial, experiments showing the bystander and genomic instability effects suggest that radiation's effects may spread considerably in cellular systems exposed to radiation. The National Academy of Science's committee on the Biological Effects of Ionizing Radiation,[16] perhaps the most influential scientific body writing on this matter, has, at the time of this writing, concluded that all of this evidence is at present not conclusive for either proof of a threshold or hormetic effect of radiation nor how the bystander and genomic instability evidence affects models predicting cancer effects at low doses.

Relying heavily on epidemiologic data from the populations discussed above, this group has concluded that the most prudent model to use at present is still the Linear, No Threshold (LNT) model.[16] This issue continues to be controversial in the scientific community, but most regulatory bodies, in the United States and elsewhere, are following this advice. Problems arise, however, when this model is used to reach *scientific* conclusions (i.e., not *operational* reasons, such as to set prudent limits on radiation dose). The public continues to believe that it has been scientifically concluded that "any dose of radiation, no matter how small" can cause cancer. Scientists, journalists, antinuclear activists, and others have published numerical estimates of the number of cancer deaths attributable or to be expected in the future from large populations of people exposed to small doses of radiation. This constitutes extrapolation of a function beyond the limits of the observed data, which every good scientist and engineer knows to be improper. Such published numerical estimates represent a misuse of the science and have served only to unnecessarily frighten the general public about low, and probably quite safe, routine exposures to radiation.

One very important point to remember is that, if radiation induces cancer in a population, the cancers are *always* expressed at some long time after the exposure (thus the early name *late effects*). This period of time between the exposure and the expression of the disease is called the *latent period*. Leukemia has the shortest latent period before expression, being as short as 5 to 10 years after the exposure. With most solid cancers, the latent period is more like 20 years. After exposure of a population to radiation, a number of erroneous efforts (whether intentionally misleading or not) have been made to show increases in cancer rates in selected populations within months after the exposure.

Use of Radiation Dosimetry in Patient Therapy Treatment Planning

Proposals for the use of patient-specific dosimetry in nuclear medicine therapy generally have been met by negative reactions in the nuclear medicine community, suggesting that:

- Performing such calculations is too difficult, requiring too much effort by the nuclear medicine staff and the patients.
- Performing such calculations is too expensive.
- There are no standardized methods for performing individualized dose calculations, and methods vary significantly among different institutions.
- Dose calculations calculated to date have had poor success in predicting tissue response.

There are good reasons for these objections, but changing conditions have provided answers to these objections, and patient-specific dose calculations in nuclear medicine therapy should now become part of routine practice. As more groups begin to gather data on different patient groups receiving therapy, dose-response relationships will begin to be better and better characterized, and success rates will increase. As image quantification methods also improve, the accuracy of dose calculations will continually increase, and physicians can have good confidence in the physicists' support of the therapy process, as is currently routine in external beam therapy.

It is not true that performing patient-specific calculations for nuclear medicine patients is too difficult or expensive to be justified. In external beam therapy, only one full or partial-body computed tomography (CT) scan is needed to provide a patient-individualized treatment plan, and three to five planar or single photon emission computed tomography (SPECT) images are needed to perform a good treatment of an individual's normal and tumor uptake and clearance. However, the patients have suffered through a battery of imaging studies, invasive and noninvasive procedures, possibly surgeries and chemotherapies, and other difficulties and personal indignations. Lying on an imaging table for 20 to 30 minutes several times is a minor inconvenience compared with other procedures they have been subjected to. The argument about cost is also not compelling. The most basic kind of dose calculation, as is performed for ^{131}I Tositumomab therapy, with the gathering of whole-body retention at multiple time points, with simple regression analysis and estimation of only "whole-body" dose, takes perhaps one-half day and costs perhaps US$200 per patient. Performing organ-based dose calculations—with three to five planar nuclear medicine scans with outlining of organ regions of interest, regression analysis of the individual organ curves, and calculation of mean organ dose using standardized dosimetry codes such as OLINDA/EXM[17]—takes perhaps 1 day of a physicist's time at a cost of perhaps US$1600 per patient. State-of-the-art individualized dosimetry—with three-dimensional dose characterization of normal organ and tumor dose, using three to five SPECT scans, image registration, Monte Carlo analysis, and characterization of dose distributions and dose-volume histograms—may take up to 3 days of intensive analysis, with an approximate cost of US$5500. This is clearly a considerable cost, but it is not at all unlike the cost currently routinely accepted for performing intensity-modulated radiation therapy (IMRT), which is also estimated to cost around US$5500 per patient.[18]

The final two objections noted above—that there are no standardized methods for performing individualized dose calculations and that dose calculations calculated to date have had poor success in predicting tissue response—deserve more attention. Dose calculations have been standardized for a number of years in the MIRDOSE[19] and OLINDA/EXM[17] personal computer codes, which implement the methods outlined by the Medical Internal Radiation Dose (MIRD) Committee of the Society of Nuclear Medicine.[20,21] These codes use the well-established standard models for reference adults,[22] children,[23] and pregnant women[24] and have been widely employed in the international nuclear medicine community. These dose calculations are very useful and nearly universally accepted in establishing standard doses for diagnostic radiopharmaceuticals for individuals of these fixed age and mass characteristics, which is needed in the drug approval process, for university approval committees to use in evaluating research proposals, and in other similar applications. Their use in therapy applications, however, has produced reproducible results, but results that have generally correlated poorly with observed effects in patient populations. Marrow toxicity from therapy with internal emitters is manifest by hematologic changes in circulating platelets, lymphocytes, granulocytes, reticulocytes, and red blood cells. Attempts to correlate hematologic toxicity with marrow dose, when marrow cells are specifically targeted, have not been particularly successful in the past, in part due to uncertainties in the actual absorbed dose, but also due to the difficulty in assessing marrow functional status prior to therapy.[25–32] Dose-response analyses showed that whole-body absorbed dose and red marrow absorbed dose are often the best predictors of hematologic toxicity, as measured by platelet toxicity grade, with red marrow dose being slightly better.

The correlations of a number of marrow toxicity indices with marrow dose for ^{90}Y Ibritumomab Tiuxetan, calculated using the reference adult phantom using the MIRDOSE code, with more than 150 subjects, were disappointing (Fig. 6.6).[33] This led to the approval of the compound with no requirement for performing patient-individualized dose calculations.

It is clear that one-dimensional dose calculations with standard reference subjects will not produce dose calculations that will be of sufficiently high quality to be used in therapy planning. Characterization of patient-specific biokinetics, though necessary, is not sufficient. Characterization of patient-specific organ mass and body anatomy must accompany the characterization of tumor and normal tissue uptake and retention. Realistic, rather than "stylized," body morphometry, based on patient images (e.g., from CT), is

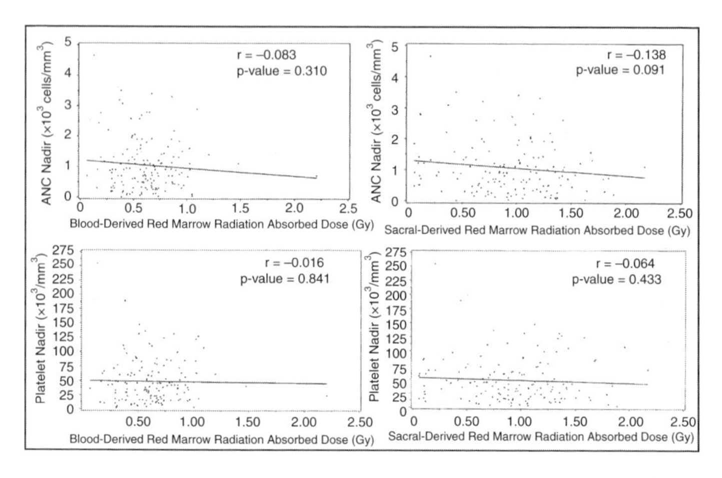

FIGURE 6.6. Correlations of neutrophil and platelet levels to radiation dose calculated with reference adult standard phantoms, for ^{90}Y Zevalin. (Reproduced by permission of the Society of Nuclear Medicine from Wiseman GA, Kornmehl E, Leigh B, Erwin WD, Podoloff DA, Spies S, Sparks RB, Stabin MG, Witzig T, White CA. Radiation dosimetry results and safety correlations from ^{90}Y-ibritumomab tiuxetan radioimmunotherapy for relapsed or refractory non-Hodgkin's lymphoma: combined data from 4 clinical trials. J Nucl Med 44:465–474, 2003.)

now possible on a patient-individualized basis and must form the basis for calculations in therapy. Furthermore, some evidence is indicating that the biologically effective dose (BED), not just the absorbed dose, is the parameter that should be characterized, in both internal and external dose calculations.[34–36] Such approaches must also be based on protocols employing imaging techniques and established data acquisition schedules that have been shown to be sufficiently detailed to produce reliable results. Such principles have been well established and recognized[37] and simply need to be defined for individual radiopharmaceutical products.

Several investigators have shown recently that patient-specific dose calculations can produce strong correlations between calculated dose and observed effects in tumors and normal tissues. The methods shown by these investigators should be widely adopted and used by others as dose calculations in nuclear medicine therapy become a routine part of providing patients with the best possible therapy and therefore the best possible and durable responses to their therapy. Shen et al.,[38] using a ^{90}Y-antibody in radioimmunotherapy, obtained an r value of 0.85 for correlation of marrow dose with observed marrow toxicity, using patient-specific marrow mass estimated from CT images and estimation of the total marrow mass from the mass of the marrow in three lumbar vertebrae (Fig. 6.7).

Siegel et al.[39] obtained a correlation coefficient of 0.86 between platelet nadir and calculated marrow dose, but with an ingenious modification based on the levels of a stimulatory cytokine (FLT3-L) measurable in peripheral blood that indicates the possible present status of a subject's marrow, in the use of an ^{131}I anti-carcinoembyronic antigen (cea) antibody (Fig. 6.8).

Whereas others have failed to find firm correlations between tumor dose and observed response, Pauwels et al. found a convincing relationship in their study of 22 patients with ^{90}Y-DOTA-Tyr3-octreotide[40] (Fig. 6.9).

Kobe et al.[41] evaluated the success of treatment of Graves' disease in 571 subjects, with the goal of delivering 250 Gy

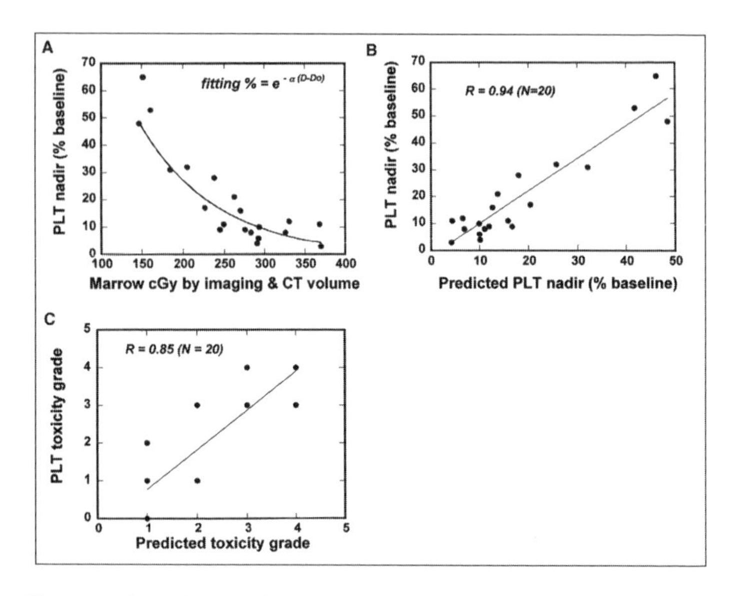

FIGURE 6.7. Correlation of platelet nadir and toxicity grade versus patient absorbed dose, with correction for patient-specific characterization of marrow mass. (Reproduced by permission of the Society of Nuclear Medicine from Shen S, Meredith RF, Duan J, Macey DJ, Khazaeli MB, Robert F, LoBuglio AF. Improved prediction of myelotoxicity using a patient-specific imaging dose estimate for non-marrow-targeting ^{90}Y-antibody therapy. J Nucl Med 43: 1245–1253, 2002.)

to the thyroid, with the end-point measure being the elimination of hyperthyroidism, evaluated 12 months after the treatment. Relief from hyperthyroidism was achieved in 96% of patients who received more than 200 Gy, even for thyroid volumes >49 mL. Individually tailored patient thyroid dosimetry was made to the targeted total dose, with ultrasound measurement of subject thyroid mass and

---

FIGURE 6.8. Correlation of platelet nadir and toxicity grade versus patient absorbed dose, with correction for patient-specific characterization of marrow status. (Reproduced by permission of the Society of Nuclear Medicine from Siegel JA, Yeldell D, Goldenberg DN,

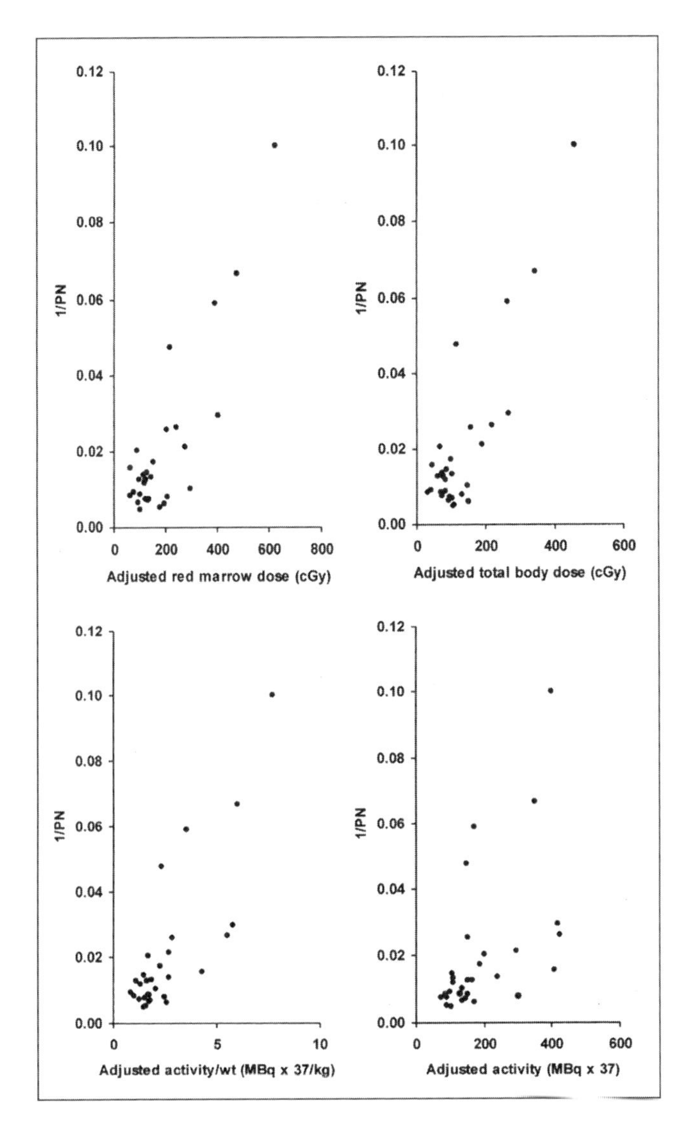

FIGURE 6.8. *(continued).* Stabin MG, Sparks RB, Sharkey RM, Brenner A, Blumenthal RD. Red marrow radiation dose adjustment using plasma FLT3-L cytokine levels: improved correlations between hematologic toxicity and bone marrow dose for radioimmunotherapy patients. J Nucl Med 44:67–76, 2003.)

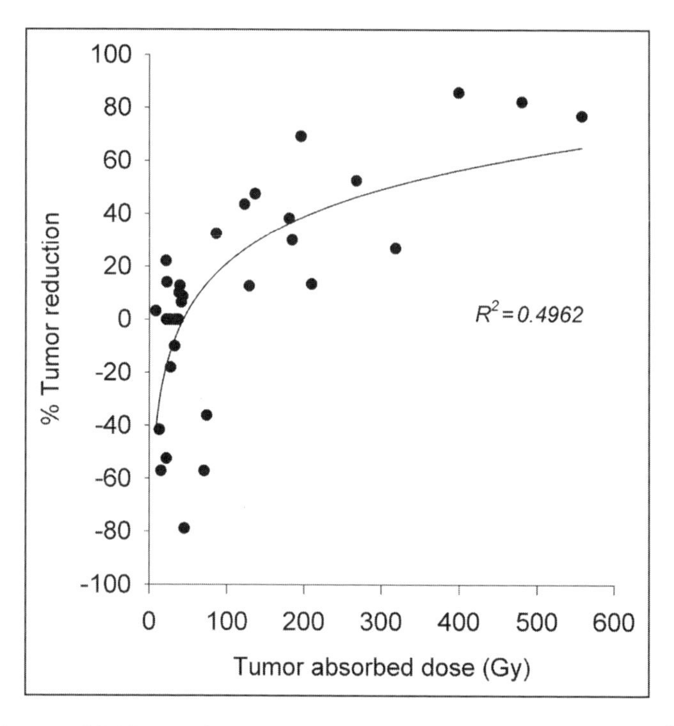

FIGURE 6.9. Tumor dose-response characterized by Pauwels et al.[40] with ^{90}Y-DOTATOC. (Reproduced by permission of the Society of Nuclear Medicine from Pauwels S, Barone R, Walrand S, Borson-Chazot F, Valkema R, Kvols LK, Krenning EP, Jamar F. Practical dosimetry of peptide receptor radionuclide therapy with ^{90}Y-labeled somatostatin analogs. J Nucl Med 46(Suppl):92S–98S, 2005.)

adjustment of the procedure to account for differences between observed effective retention half-times between studies involving the tracer activity and the therapy administration. These authors note that success rates with more traditional treatments (not using individually tailored dosimetry) are typically at best 60% to 80%.

In conclusion, then, our understanding of radiation biology from internal emitters requires considerable attention in the years to come. Improving this understanding can only come if careful dosimetry is performed with many therapy patients, as dose-effect relationships cannot be studied at all if there is no calculation of dose. Providing better and more durable outcomes for cancer patients requires more aggressive and optimized therapy, which again is not possible in careful and accurate dosimetry. A paradigm shift is needed in the nuclear medicine clinic to accommodate these changes and improve patient therapy.

References

1. Mullner R. Deadly Glow. The Radium Dial Worker Tragedy. American Public Health Association, Washington, DC, 1989.
2. Pimblott SM, LaVerne JA. Stochastic simulation of the electron radiolysis of water and aqueous solutions. J Phys Chem A 101:5828–5838, 1977.
3. Becker D, Sevilla MD, Wang W, LaVere T. The role of waters of hydration in direct-effect radiation damage to DNA. Radiat Res 148:508–510, 1997.
4. Wright HA, Magee JL, Hamm RN, Chatterjee A, Turner JE, Klots CE. Calculations of physical and chemical reactions produced in irradiated water containing DNA. Radiat Prot Dosimetry 13:133–136, 1985.
5. Turner JE, Hamm RN, Ritchie RH, Bolch WE. Monte Carlo track-structure calculations for aqueous solutions containing biomolecules. Basic Life Sci 63:155–66, 1994.
6. Bergonie J, Tribondeau L. De quelques resultats de la Radiotherapie, et esaie de fixation d'une technique rationelle. Comptes Rendu des Seances de l'Academie des Sciences 143:983–985, 1906.
7. Available at http://www.rerf.or.jp/index_e.html.
8. Luckey TD. Radiation Hormesis. CRC Press, Boca Raton, FL, 1991.
9. Jonah CD, Miller JR. Yield and decay of the OH radical from 100 ps to 3 ns. J Phys Chem 81:1974–1976, 1977.
10. NCRP. The Relative Biological Effectiveness of Radiations of Different Quality. NCRP Report No. 104. National Council on Radiation Protection and Measurements, Bethesda, MD, 1990.
11. Humm JL, Howell RW, Rao DV. AAPM Report No. 49, Dosimetry of Auger-Electron-Emitting Radionuclides. Med Phys 21(12), 1994.

12. Brooks AL. Evidence for "bystander effects" in vivo. Human Exp Toxicol 23:67–70, 2004.
13. Hall EJ. The bystander effect. Health Phys 85:31–35, 2003.
14. Morgan WF. Non-targeted and delayed effects of exposure to ionizing radiation: II. Radiation-induced genomic instability and bystander effects in vivo, clastogenic factors and transgenerational effects. Radiat Res 159:581–596, 2003.
15. Kishikawa H, Wang K, Adelstein SJ, Kassis AI. Inhibitory and stimulatory bystander effects are differentially induced by iodine-125 and iodine-123. Radiat Res 165:688–694, 2006.
16. National Academy of Sciences. Health Risks from Exposure to Low Levels of Ionizing Radiation: BEIR VII Phase 2. The National Academies Press, Washington, DC, 2006.
17. Stabin MG, Sparks RB, Crowe E. OLINDA/EXM: the second-generation personal computer software for internal dose assessment in nuclear medicine. J Nucl Med 46:1023–1027, 2005.
18. Flux G, The Royal Marsden. Cost estimates: personal communication, 2004.
19. Stabin M. MIRDOSE - the personal computer software for use in internal dose assessment in nuclear medicine. J Nucl Med 37:538–546, 1996.
20. Loevinger R, Budinger T, Watson E. MIRD Primer for Absorbed Dose Calculations. Society of Nuclear Medicine, New York, 1988.
21. Snyder W, Ford M, Warner G, Watson S. "S," absorbed dose per unit cumulated activity for selected radionuclides and organs. MIRD Pamphlet No. 11. Society of Nuclear Medicine, New York, 1975.
22. Snyder W, Ford M, Warner G. Estimates of specific absorbed fractions for photon sources uniformly distributed in various organs of a heterogeneous phantom. MIRD Pamphlet No. 5, revised. Society of Nuclear Medicine, New York, 1978.
23. Cristy M, Eckerman K. Specific absorbed fractions of energy at various ages from internal photons sources. ORNL/TM-8381 V1-V7. Oak Ridge National Laboratory, Oak Ridge, TN, 1987.
24. Stabin M, Watson E, Cristy M, Ryman J, Eckerman K, Davis J, Marshall D, Gehlen K. Mathematical models and specific absorbed fractions of photon energy in the nonpregnant adult female and at the end of each trimester of pregnancy. ORNL Report ORNL/TM-12907. Oak Ridge National Laboratory, Oak Ridge, TN, 1995.

25. DeNardo DA, DeNardo GL, O'Donnell RT, et al. Imaging for improved prediction of myelotoxicity after radioimmunotherapy. Cancer 80:2558–2566, 1997.
26. Siegel JA, Lee RE, Pawlyk DA, et al. Sacral scintigraphy for bone marrow dosimetry in radioimmunotherapy. Int. J Rad Appl Instrum B16:553–559, 1989.
27. Siegel JA, Wessels, Watson EE, et al. Bone marrow dosimetry and toxicity for radioimmunotherapy. Antibody Immunoconj Radiopharmacol 3:213–233, 1990.
28. Lim S-M, DeNardo GL, DeNardo DA et al. Prediction of myelotoxicity using radiation doses to marrow from body, blood and marrow sources. J Nucl Med 38:1474–1378, 1997.
29. Breitz H, Fisher D, Wessels B. Marrow toxicity and radiation absorbed dose estimates from rhenium-186-labeled monocolonal antibody. J Nucl Med 39:1746–1751, 1998.
30. Eary JF, Krohn KA, Press OWQ, et al. Importance of pretreatment radiation absorbed dose estimation for radioimmunotherapy of non Hodgkin's lymphoma. Nucl Med Biol 24:635–638, 1997.
31. Behr TM, Sharkey RM, Juweid ME, Dunn RM, Vgg RC, Siegel JA, Goldenberg DM. Hematological toxicity in the radioimmunotherapy of solid cancers with ^{131}I-labeled anti-CEA NP-4 IgG1: dependence on red marrow dosimetry and pretreatment. In: Sixth International Radiopharmaceutical Dosimetry Symposium. Vol I. Stelson A, Stabin M, Sparks R, eds, ORAU, 1999, pp. 113–125.
32. Juweid ME, Zhang C-H, Blumenthal RD, Hajjar G, Sharkey RM, Goldenberg DM. Prediction of hematologic toxicity after radioimmunotherapy with ^{131}I-labeled anticarcinoembryoinic antigen monoclonal antibodies. J Nucl Med 10:1609–1616, 1999.
33. Wiseman GA, Kornmehl E, Leigh B, Erwin WD, Podoloff DA, Spies S, Sparks RB, Stabin MG, Witzig T, White CA. Radiation dosimetry results and safety correlations from ^{90}Y-ibritumomab tiuxetan radioimmunotherapy for relapsed or refractory non-Hodgkin's lymphoma: combined data from 4 clinical trials. J Nucl Med 44:465–474, 2003.
34. Dale R, Carabc-Fernandez A. The radiobiology of conventional radiotherapy and its application to radionuclide therapy. Cancer Biother Radiopharm 20:47–51, 2005.
35. Bodey RK, Flux GD, Evans PM. Combining dosimetry for targeted radionuclide and external beam therapies using the biologically effective dose. Cancer Biother Radiopharm 18: 89–97, 2003.

36. Barone R, Borson-Chazot F, Valkema R, Walrand S, Chauvin F, Gogou L,. Kvols LK, Krenning EP, Jamar F, Pauwels S. Patient-specific dosimetry in predicting renal toxicity with ^{90}Y-DOTATOC: relevance of kidney volume and dose rate in finding a dose-effect relationship. J Nucl Med 46: 99S–106S, 2005.
37. Siegel J, Thomas S, Stubbs J, Stabin M, Hays M, Koral K, Robertson J, Howell R, Wessels B, Fisher D, Weber D, Brill A. MIRD Pamphlet No. 16: techniques for quantitative radiopharmaceutical biodistribution data acquisition and analysis for use in human radiation dose estimates. J Nucl Med 40: 37S–61S, 1999.
38. Shen S, Meredith RF, Duan J, Macey DJ, Khazaeli MB, Robert F, LoBuglio AF. Improved prediction of myelotoxicity using a patient-specific imaging dose estimate for non-marrow-targeting ^{90}Y-antibody therapy. J Nucl Med 43:1245–1253, 2002.
39. Siegel JA, Yeldell D, Goldenberg DN, Stabin MG, Sparks RB, Sharkey RM, Brenner A, Blumenthal RD. Red marrow radiation dose adjustment using plasma FLT3-L cytokine levels: improved correlations between hematologic toxicity and bone marrow dose for radioimmunotherapy patients. J Nucl Med 44:67–76, 2003.
40. Pauwels S, Barone R, Walrand S, Borson-Chazot F, Valkema R, Kvols LK, Krenning EP, Jamar F. Practical dosimetry of peptide receptor radionuclide therapy with ^{90}Y-labeled somatostatin analogs. J Nucl Med 46(Suppl):92S–98S, 2005.
41. Kobe C, Eschner W, Sudbrock F, Weber I, Marx K, Dietlein M, Schicha H. Graves' disease and radioiodine therapy: is success of ablation dependent on the achieved dose above 200 Gy? Nuklearmedizin 2007 (in press).

7 Regulatory Aspects of Dose Calculations*

The Philosophy of Radiation Protection

The practice of radiation protection involves three fundamental principles:

1. Justification: No practice should be undertaken unless sufficient benefit to the exposed individuals will offset the radiation detriment.
2. Optimization: The magnitude of individual doses, the number of people exposed, and the likelihood of incurring exposures should be kept as low as reasonably achievable (ALARA), economic and social factors being taken into account.
3. Limitation: The exposure of individuals should be subject to dose limits. These limits are designed to prevent deterministic effects and to reduce stochastic effects to an "acceptable" level. [1]

Radiation protection is managed through an interaction between established regulatory bodies, scientific advisory bodies, and users of these technologies. The scientific advisory bodies were formed very early in the history of the use of radiation and continue to function today. They have no

*Portions of this chapter are reproduced and adapted, with permission, from Stabin, M. Radiation Protection and Dosimetry. Chapter 7: The basis for regulation of radiation exposure. Springer, New York, 2007.

"official" status, generally speaking. Some are appointed by a parent organization (e.g., the International Atomic Energy Agency was chartered in July 1957 by the United Nations), and others were formed as people perceived the necessity for them to exist, and their existence continues as long as some source of funding exists and there is a continued perceived need for their input. Some came into existence and were eliminated or replaced by other bodies over time. Scientific advisory bodies do not have authority to issue or enforce regulations. However, their recommendations often serve as the basis for the radiation protection regulations adopted by the regulatory authorities in the United States and most other nations. Much of the material in the following sections is derived from information in Lauriston Taylor's notable compendium on the subject[2] and another published work by this author.[3]

Scientific Advisory Bodies

The International Commission on Radiological Protection and the National Council on Radiation Protection and Measurements

In 1928, at the Second International Congress of Radiology meeting in Stockholm, Sweden, the first radiation protection commission was created. The body was named the International X-Ray and Radium Protection Commission (ICXRP). It was charged with developing recommendations concerning radiation protection. In 1950, to better reflect its role in a changing world, the commission was reorganized and renamed the International Commission on Radiological Protection (ICRP). The ICRP is still very active today and is considered to be the leading organization that develops recommendations for radiation protection, many of which are intended to (and do) influence the regulatory process in most countries. In 1929, the U.S. Advisory Committee on X-Ray and Radium Protection (ACXRP) was formed.

In 1964, the committee was congressionally chartered as the National Council on Radiation Protection and Measurements (NCRP). Both the NCRP and ICRP put out scientific documents that discuss the state of knowledge in a particular area of radiation protection science or put forth new knowledge or recommendations for practice.

The International Atomic Energy Agency

The International Atomic Energy Agency (IAEA) was chartered in July 1957 as an autonomous intergovernmental organization by the United Nations (UN). The IAEA gives advice and technical assistance to UN Member States on nuclear power development, health and safety issues, radioactive waste management, and on a broad range of other areas related to the use of radioactive material and atomic energy in industry and government. As government bodies do not necessarily have to adopt the recommendations of the ICRP and NCRP, UN Member States do not have to follow IAEA recommendations. If they chose to ignore the IAEA recommendations, however, funding for international programs dealing with the safe use of atomic energy and radioactive materials can be withheld, and, in matters related to safeguarding special nuclear material, UN resolutions may be enforced legally, using government and even military intervention if needed. Many of the IAEA recommendations follow ICRP recommendations on radiation protection philosophy and numerical criteria. The IAEA has published a number of useful scientific documents, including tables of dose values for workers and the public, mostly drawing on results generated by the ICRP. The IAEA sponsors much international research in many areas of radiation research, with funds mostly going to developing countries.

The National Academy of Sciences

The National Academy of Sciences (NAS) is a private, non-profit, self-perpetuating society of distinguished scholars engaged in scientific and engineering research, dedicated to the furtherance of science and technology and to their use for the general welfare. Upon the authority of the charter granted to it by Congress in 1863, the academy has a mandate that requires it to advise the federal government on scientific and technical matters. Members and foreign associates of the academy are elected in recognition of their distinguished and continuing achievements in original research; election to the academy is considered one of the highest honors that can be accorded a scientist or engineer. The academy membership comprises approximately 1900 members and 300 foreign associates, of whom more than 170 have won Nobel Prizes. The NAS publishes on a variety of topics. Its most influential works in the area of radiation protection are its summaries of the Biological Effects of Ionizing Radiation (BEIR). BEIR I, BEIR III, and BEIR V have all been highly influential in the setting of radiation protection standards, based on the total knowledge of radiation effects in humans and animals. BEIR VII was recently released, and the conclusions were relatively controversial, due to the current debates about low levels of radiation and health effects. The authors endorsed again the use of a linear, no threshold (LNT) model for prediction of radiation carcinogenesis at low doses and dose rates. Some modification to specific model results were presented, based on new cancer data and new dosimetry analyses from the Hiroshima and Nagasaki bombings, but the ultimate conclusions were basically in agreement with those given in BEIR V. (Note: BEIR VI is a report strictly on dose-effect relationships for radon.)

The United Nations Scientific Committee on the Effects of Atomic Radiation

In 1955, the General Assembly of the United Nations established the United Nations Scientific Committee on the Effects of Atomic Radiation (UNSCEAR) in response to widespread concerns regarding the effects of radiation on human health and the environment. At that time, nuclear weapons were being tested in the atmosphere, and radioactive debris was dispersing throughout the environment, reaching the human body through intake of air, water, and foods. The committee was requested to collect, assemble, and evaluate information on the levels of ionizing radiation and radionuclides from all sources (natural and produced by man) and to study their possible effects on man and the environment. The committee consists of scientists from 21 Member States. These member states are Argentina, Australia, Belgium, Brazil, Canada, China, Egypt, France, Germany, India, Indonesia, Japan, Mexico, Peru, Poland, Russia, Slovakia, Sudan, Sweden, United Kingdom, and the United States of America. The UNSCEAR secretariat, which gives the committee the necessary assistance in carrying out its work, is located in Vienna; it consults with scientists throughout the world in establishing databases of exposures and information on the effects of radiation. The committee produces the UNSCEAR Reports, which are detailed reports to the General Assembly. The most influential of those are the reports on the Sources and Effects of Ionizing Radiation, which catalogue human exposure to natural background and to occupational and medical radiation sources worldwide.

The International Commission on Radiation Units and Measurements

The International Commission on Radiation Units and Measurements (ICRU) was established in 1925 by the International Congress of Radiology. Since its inception, it has had as its principal objective the development of internationally acceptable recommendations regarding (1) quantities and units of radiation and radioactivity; (2) procedures suitable for the measurement and application of these quantities in diagnostic radiology, radiation therapy, radiation biology, and industrial operations; (3) physical data needed in the application of these procedures, the use of which tends to assure uniformity in reporting. The ICRU endeavors to collect and evaluate the latest data and information pertinent to the problems of radiation measurement and dosimetry and to recommend in its publications the most acceptable values and techniques for current use. The ICRU has published a series of useful documents, most importantly defining radiation quantities and units, but also discussing the state of the science in various applications of radiation in general protection and medicine.

Regulatory Bodies

The U.S. Food and Drug Administration

The U.S. Food and Drug Administration (FDA) sets standards for the use of lasers (21CFR) and other non-ionizing radiation, food irradiation, and pharmaceuticals. Medical imaging agents are submitted for approval in:

- Investigational new drug applications (INDs)
- New drug applications (NDAs)
- Biologics license applications (BLAs)
- Abbreviated NDAs (ANDAs)
- Supplements to NDAs or BLAs.

Radiation safety assessment associated with the approval of use of medical imaging agents shall follow these criteria[4]:

- … [shall] allow a reasonable calculation of the radiation absorbed dose to the whole body and to critical organs upon administration to a human subject …
- At a minimum, … radiation absorbed dose estimates [shall] be provided for all organs and tissues in the standardized anthropomorphic phantoms established in the literature …
- For diagnostic radiopharmaceuticals … [one should calculate] the *effective dose* as defined by the International Commission on Radiological Protection (ICRP) in its ICRP Publication 60 (this quantity is not meaningful for therapeutic radiopharmaceuticals) …
- The amount of the radiation absorbed dose delivered by internal administration of diagnostic radiopharmaceuticals be calculated by standardized methods [should be provided] …

- The methodology used to assess radiation safety [should] be specified including reference to the body models that were used ...
- The mathematical equations used to derive the time activity curves and the radiation absorbed dose estimates [should] be provided along with a full description of assumptions that were made ...
- Sample calculations and all pertinent assumptions [should] be listed and submitted ...
- The reference to the body, organ, or tissue model used in the dosimetry calculations [should] be specified, particularly for new models being tested. If a software program was used to calculate the radiation doses ...
- [one should provide]
  - a full description of the code, including official name, version number, and computing platform;
  - a literature citation for the code; and
  - photocopies of the code's output, preferably showing all of the user input data and model choices."[†]

An assessment of any significant radiation hazards to other patients and health care workers should also be undertaken. Applicants should provide a description of which organs have a significant accumulation of activity over time, information on the activity levels at different times (with at least two time points obtained per phase of radionuclide uptake or clearance), and an evaluation of the time integrals of activity, description of how they were obtained, and a description of how they were combined with dose conversion factors to obtain doses (if not done by software as noted earlier).

Approval of a new medical imaging agent includes several phases:

- A *preclinical phase*, in which studies in an appropriate animal species are carefully planned and executed, to provide a preliminary assessment of the possible radiation doses expected in human subjects. As noted in previous

[†] In this quoted section, ellipses and brackets are inserted for readability. Excerpt from U.S. Food and Drug Administration. Guidance for Industry Developing Medical Imaging Drug and Biological Products, Part 1: Conducting Safety Assessments. U.S. FDA, Washington, DC, 2004.

chapters, extrapolation of animal data to humans is far from an exact science. Results from such studies represent an important first step in the evaluation, but they should always be viewed with caution, in anticipation of more reliable results from the human data obtained in other phases.

- *Phase 1 studies* of medical imaging agents, which are designed to obtain pharmacokinetic and human safety assessments, based on a single mass administration and escalating mass administrations of the drug or biological product. The FDA recommends that evaluation of medical imaging agents that target a specific metabolic process or receptor include assessments of its potential effects on any relevant processes or receptors.
- *Phase 2 studies* of medical imaging agents include[5]:
  - "refining the agent's clinically useful mass dose and radiation dose ranges or dosage regimen (e.g., bolus administration or infusion) in preparation for phase 3 studies;
  - answering outstanding pharmacokinetic and pharmacodynamic questions;
  - providing preliminary evidence of efficacy and expanding the safety database;
  - optimizing the techniques and timing of image acquisition;
  - developing methods and criteria by which images will be evaluated;
  - evaluating other critical questions about the medical imaging agent."[‡]
- *Phase 3 studies* are designed to confirm the principal hypotheses developed in earlier studies, demonstrating the efficacy of the compound and method employed, to verify the safety of the use of the medical imaging agent, and to validate the necessary instructions for use of the compound and for imaging in the population for which the agent is intended.

[‡] Excerpted from U.S. Food and Drug Administration. Guidance for Industry Developing Medical Imaging Drug and Biological Products Part 3: Design, Analysis, and Interpretation of Clinical Studies. U.S. FDA, Washington, DC, 2004.

Much more detail is available about the appropriate conduct of such trials and submission of documentation to the FDA in support of submissions for new agents; see Refs. 1 and 2. These documents are currently available on the FDA Web site in electronic form.

In addition, the FDA permits basic research using radioactive drugs in humans without an IND when the compound is administered under certain conditions, defined in the radioactive drug research committee (RDRC) program.[6] The conditions are that:

- The research is considered basic science research and is done for the purpose of advancing scientific knowledge, for example, research intended to obtain basic information regarding the metabolism (including kinetics, distribution, dosimetry, and localization) of a radioactive drug or involving human physiology, pathophysiology, or biochemistry.
- The drug must not be intended for immediate therapeutic, diagnostic, or similar purposes, and the study is not intended to determine the safety and effectiveness of the drug in humans.

An RDRC approved research study must have the following components:

- Recognized and qualified study investigators.
- A properly licensed medical facility that will possess and handle radioactive materials.
- Careful and approved selection and consent of research subjects.
- A quality assurance program for the compound administered.
- An approved research protocol design.
- A system for reporting of adverse events to the institution's RDRC.
- Prior approval by the institution's institutional review board (IRB).
- Assurance that the pharmacologic dose of the radioactive drug to be administered is not known to cause any clinically detectable pharmacologic effect in humans.

- Verification that the radiation dose to be administered is justified by the quality of the study being undertaken and the importance of the information it seeks to obtain and is within the radiation dose limits specified in 21CFR361.1(b)(3).

The dose limits specified in 21CFR361 are as follows:

Radiation dose to an adult research subject from a single study or cumulatively from a number of studies conducted within 1 year:

Whole body, active blood-forming organs, lens of the eye, and gonads

Single dose	30 mSv (3 rem)
Annual and total dose commitment	50 mSv (5 rem)

Other organs

Single dose	50 mSv (5 rem)
Annual and total dose commitment	150 mSv (15 rem)

Research subjects under 18 years of age shall not receive doses exceeding 10% of those for adults. The dose calculations must consider all radioactive material in the product, including significant contaminants and/or impurities. Radiation doses from other procedures involving ionization radiation that are part of the research study (i.e., would not have been received by the subject except due to their participation in the study) must be included in the total dose received and compared with the above limits. It should be noted these dose limitations may restrict the number of hybrid imaging studies (PET/CT or SPECT/CT) that can be performed under an RDRC protocol.

The U.S. Nuclear Regulatory Commission

The Atomic Energy Act (AEA) of 1954[7] is the fundamental U.S. law regulating both the civilian and the military uses of nuclear materials. Under the Atomic Energy Act of 1954, a single agency, the Atomic Energy Commission, had responsibility for the development and production of nuclear weapons and for both the development and the safety regulation of the civilian uses of nuclear materials. On the civilian side, it provides for both the development and the regulation of the uses of nuclear materials and facilities in the United States, declaring the policy that "the development, use, and control of atomic energy shall be directed so as to promote world peace, improve the general welfare, increase the standard of living, and strengthen free competition in private enterprise."[7] The AEA requires that civilian uses of nuclear materials and facilities be licensed, and it *empowered the AEC* to establish by rule or order and to enforce such standards to govern these uses as "the Commission may deem necessary or desirable in order to protect health and safety and minimize danger to life or property."[7] As we will see later, the two functions (civilian and military) were later separated, and the civilian portion was given to the (then formed) Nuclear Regulatory Commission (NRC), which still functions today.

Agreement State Concept

Under section 274 of the AEA, *the NRC may enter into an agreement with a state* for discontinuance of the NRC's regulatory authority over some materials licensees within the state. States already regulate:

1. Naturally occurring radioactive materials (NORMs).
2. Radiation-producing machines (medical and industrial X-rays, particle accelerators).
3. Radioactivity produced in accelerators.

To become an Agreement State, the state must first show that its regulatory program is compatible with the NRC's and adequate to protect public health and safety.[8] The NRC retains authority over nuclear power plants, but the Agreement State then is given power to regulate within its borders the use of:

1. by-product
2. source
3. special nuclear material (in small quantities)

By-product material is (1) any radioactive material (except special nuclear material) yielded in or made radioactive by exposure to the radiation incident to the process of producing or using special nuclear material (as in a reactor); (2) the tailings or wastes produced by the extraction or concentration of uranium or thorium from ore.

Source material is uranium or thorium, or any combination thereof, in any physical or chemical form or ores that contain by weight one-twentieth of 1% (0.05%) or more of (1) uranium, (2) thorium, or (3) any combination thereof. Source material does not include special nuclear material.

Special nuclear material is plutonium, uranium-233, or uranium enriched in the isotopes uranium-233 or uranium-235.

The NRC periodically assesses the compatibility and adequacy of the state's program for consistency with the national program. Listed below are those states that currently are Agreement States:

Alabama	Kansas	New York
Arizona	Kentucky	North Carolina
Arkansas	Louisiana	North Dakota
California	Maine	Oregon
Colorado	Maryland	Rhode Island
Florida	Mississippi	South Carolina
Georgia	Nebraska	Tennessee
Illinois	Nevada	Texas
Iowa	New Hampshire	Utah
New Mexico	Washington	

Low-Level Waste Disposal

Hospitals and other medical facilities generate large amounts of low-level waste. The Low-Level Radioactive Waste Policy Amendments Act of 1985 (LLWPA) gave states the responsibility to dispose of low-level radioactive waste generated within their borders and allows them to form compacts to locate facilities to serve a group of states. The act provides that the facilities will be regulated by the NRC or by states that have entered into agreements with the NRC under section 274 of the Atomic Energy Act. The State Compact for LLW is a complete mess at the moment (Fig. 7.1). There was once a nice scheme, dividing the country into several large regions within which a site would be selected. This rapidly disintegrated into a political quagmire.

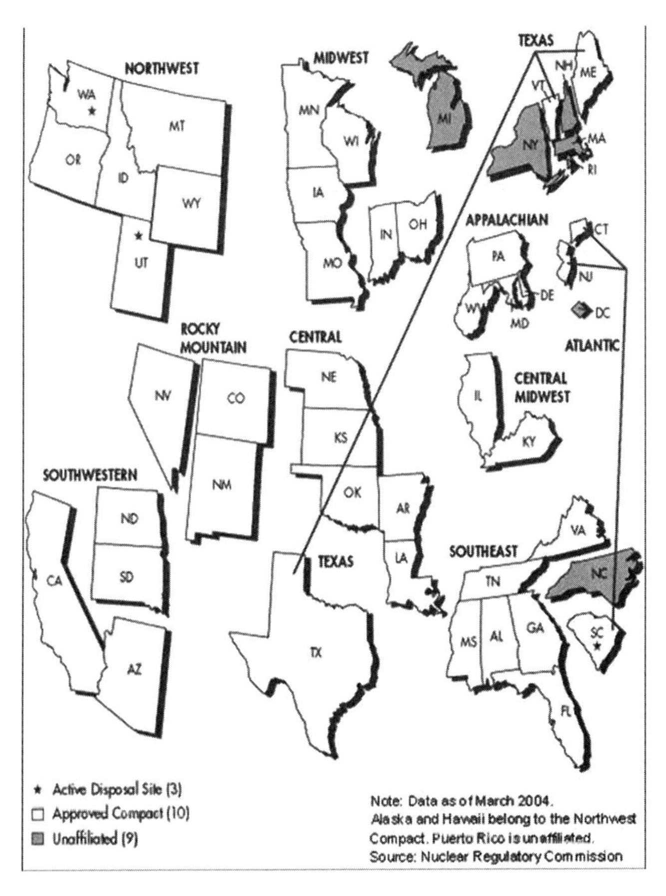

FIGURE 7.1. Current status of potential low-level waste compacts in the United States. Data as of March 2004. Alaska and Hawaii belong to the Northwest Compact. Puerto Rico is unaffiliated. (From http://www.nrc.gov/waste/llw-disposal/compacts.html.)

At present, there are three operational LLW disposal facilities, but only two—one at Richland, Washington, near Hanford; and one at Barnwell, South Carolina, near Savanna River—are open to a wide variety of LLWs. A third facility at Clive, Utah, accepts a few limited categories of LLW. The future of other sites in North Carolina, Texas, California, and elsewhere is in question. When a site is chosen, an orchestrated protest is engaged by organized antinuclear groups, which generally is able to cause enough concern in the public to delay or stall the process of approval. The focus of many of these groups is their opposition to nuclear power production. Most of these orchestrated protests involve emotionally charged and often disingenuous arguments to try and frighten the general public about low-level exposures to radiation and thus engender organized opposition to the siting of any disposal facility. As our understanding of the risks of exposure to low levels of radiation is still fraught with many uncertainties (see Chapter 6), these groups can pick and choose the research findings that further their points, while ignoring others, and emphasize uncertainties, which cause anxiety in the minds of the public where the subject of cancer is concerned. It is thus much easier to add energy to the "not in my backyard" (NIMBY) sentiment than it is to overcome such sentiments with reason and data, as proponents attempt to do. The antinuclear groups often do not realize, however, the potential impact their activities have on the practice of nuclear medicine, rather than only on nuclear power.

What exactly is "low-level waste"? Table 7.1 provides the operational definitions of different types of nuclear waste. Waste is generally defined in categories based on its *origins*, not necessarily its *present hazard level.* A high-activity ^{137}Cs source, definitely capable of delivering high doses if contacted, will be a type of "low-level" waste.

10CFR20

Title 10 of the U.S. Code of Federal Regulations, Part 20 (10CFR20), is the main piece of legislation that governs radiation worker exposures. Other regulations may cover environmental releases, transportation of radioactive

TABLE 7.1. Categories of radioactive waste and their definitions.

Category of radioactive waste	Definition
High-level waste (HLW)	1. *Spent fuel:* irradiated commercial reactor fuel. 2. *Reprocessing waste:* liquid waste from solvent extraction cycles in reprocessing. Also the solids into which liquid wastes may have been converted. NOTE: The Department of Energy defines HLW as reprocessing waste only, whereas the Nuclear Regulatory Commission defines HLW as spent fuel *and* reprocessing waste.
Transuranic waste (TRU)	Waste containing elements with atomic numbers (number of protons) greater than 92, the atomic number of uranium. (Thus the term *transuranic*, or "above uranium.") TRU includes only waste material that contains transuranic elements with half-lives greater than 20 years and concentrations greater than 100 nanocuries per gram. If the concentrations of the half-lives are below the limits, it is possible for waste to have transuranic elements but not be classified as TRU waste.
Low-level waste (LLW)	Defined by what it is not. It is radioactive waste not classified as high-level, spent fuel, transuranic or by-product material such as uranium mill tailings. LLW has four subcategories: Classes A, B, C, and Greater-Than Class-C (GTCC). On average, Class A is the least hazardous and GTCC is the most hazardous.

Source: Reproduced with permission from Stabin, M. Radiation Protection and Dosimetry. Chapter 7: The basis for regulation of radiation exposure. Springer, New York, 2007. Data from http://www.nrc.gov/reading-rm/doc-collections/cfr/part035/; http://www.nrc.gov/about-nrc/state-tribal/agreement-states.html.

materials, and other issues, but the most important code is 10CFR20. Some of the primary dose limits of interest for medical facilities are

1. The annual limit for radiation workers, which is the more limiting:

   (i) The total effective dose equivalent being equal to 0.05 Sv (5 rem); or

(ii) The sum of the deep-dose equivalent and the committed dose equivalent to any individual organ or tissue other than the lens of the eye being equal to 0.5 Sv (50 rem).

2. The annual limits to the lens of the eye, to the skin of the whole body, and to the skin of the extremities, which are
   (i) A lens dose equivalent of 0.15 Sv (15 rem), and
   (ii) A shallow-dose equivalent of 0.5 Sv (50 rem) to the skin of the whole body or to the skin of any extremity.
3. The licensee shall ensure that the dose equivalent to the *embryo/fetus* during the entire pregnancy, due to the occupational exposure of a *declared pregnant woman*, does not exceed 5 mSv (0.5 rem).
4. The total effective dose equivalent to individual *members of the public* from the licensed operation does not exceed 1 mSv (0.1 rem) in a year, exclusive of the dose contributions from background radiation, from any medical administration the individual has received.

Dose from Radioactive Patients Released After Nuclear Medicine Therapy

Patients who receive therapeutic amounts of radiopharmaceuticals are a potentially significant source of radiation to their family members, members of the public whom they pass by on their way from the hospital to their homes, and others. For many decades, the release criterion for such patients was primarily activity, based on patients treated with ^{131}I sodium iodide, used to treat hyperthyroidism (Graves' disease) or thyroid cancer, as this comprised almost all radiation therapy that involved radioactive material with a significant gamma component. The release limit (which no one knows how it was originally derived[9]) was that patients could be let go when their activity level was 1100 MBq (30 mCi), or the dose rate at 1 m from the patient was 50 μSv/h (5 mrem/h). In a new version of 10CFR35.75, issued in 1997, the NRC changed the system to be more objectively based on a purely dose-based criterion and to cover the many more therapeutic radiopharmaceuticals in use. Licensees are now able to release patients regardless of how much administered activity they received, if the radiation dose to any individual from exposure to the released patient can be shown to be less than 5 mSv (0.5 rem), integrated over all time after patient release. The rule states that the "licensee shall provide the released individual, or the individual's parent or guardian, with instructions, including written instruction, on actions recommended to maintain doses to other individuals as low as is reasonably achievable. . . ."[10] The NRC did not intend to enforce patient compliance with the instructions nor is it the licensee's responsibility to do so. But hospitals do need to keep records showing that they have ascertained that the doses to the maximally exposed individual is "not likely" to be above the stated dose limit of 5 mSv. The NRC published a regulatory guide, NRC Regulatory Guide 8.39; these documents do not carry any force of law, *unless the licensee formally adopts them in his or her license* as part of the facility's official procedures. The NRC has formally noted that other good methods can be used for these calculations and may be accepted by the commission if they can be shown to be sound. The method used in the regulatory guide was quite conservative in a number of aspects. First, the patient was treated as a point source in calculation of external exposure rates. As can be noted by solving the equations in the early parts of this chapter, lower doses will be delivered from line orvolume sources than from point sources. Patients will have activity distributed throughout their entire bodies, and some self-attenuation will occur; thus, the use of a point source is quite conservative. Then, the decay of activity was assumed to be only by physical decay of the radionuclide; biological elimination by the patient was not considered. Actual measurements on patients' family members by one group of authors indeed showed that the real doses received by people are significantly less than that assumed by the calculation methods in the regulatory guide.[11] The equation used was:

$$D(\infty) = \frac{34.6 \Gamma Q_0 T_{\mathrm{p}} \mathrm{OF}}{r^2}$$

Here, $D(\infty)$ is the dose integrated over all time, Γ is the radionuclide specific gamma constant, Q_0 is patient activity at time of release, T_p is the radionuclide half-life, OF is the assumed occupancy factor, and r is the assumed average distance from a subject over the time of irradiation. For short-lived nuclides ($T_p \leq 1$ d), an OF of 1.0 was used, and for others, an OF of 0.25 was used. The default average distance from a subject was assumed to be 1 m. The NRC provided a default table of activity levels and dose rates for various radionuclides at which they deem the dose criterion will be met. Table 7.2 contains a sample portion of that table.

TABLE 7.2. Activities and dose rates recommended in NRC Regulatory Guide 8.39 for patient release.

	Activity at or below which patients may be released		Dose rate at 1 m at or below which patients may be released	
Radionuclide	GBq	mCi	mSv/h	mrem/h
^{111}Ag	19	520	0.08	8
^{198}Au	3.5	93	0.21	21
^{51}Cr	4.8	130	0.02	2
^{64}Cu	8.4	230	0.27	27
^{67}Cu	14	390	0.22	22
^{67}Ga	8.7	240	0.18	18
^{123}I	6.0	160	0.26	26
^{125}I	0.25	7	0.01	1
^{125}I implant	0.33	9	0.01	1
^{131}I	1.2	33	0.07	7
^{111}In	2.4	64	0.2	20
^{192}Ir implant	0.074	2	0.008	0.8
^{103}Pd implant	1.5	40	0.03	3
^{186}Re	28	770	0.15	15
^{188}Re	29	790	0.20	20
^{47}Sc	11	310	0.17	17
^{75}Se	0.089	2	0.005	0.5
^{153}Sm	26	700	0.3	30
^{89}Sr	1.1	29	0.04	4
^{99m}Tc	28	760	0.58	58
^{201}Tl	16	430	0.19	19
^{169}Yb	0.37	10	0.02	2

Source: Adapted from U.S. Nuclear Regulatory Commission. Regulatory Guide 8.39. Release of Patients Administered Radioactive Materials. U.S. Nuclear Regulatory Commission, Office of Nuclear Regulatory Research, Washington, DC, 1997.

These release criteria are clearly an improvement over the prior situation in which most releases were somehow tied to the one simple activity limit. However, the method has some shortcomings, for example being overly conservative in the use of a point source model and not accounting for biologic removal of the radiopharmaceuticals. The calculations have been shown to be conservative,[11–13] and several authors have called for a more careful evaluation and reissuance of the guidelines.

The U.S. Department of Transportation

The U.S. Department of Transportation (DOT) regulates the shipment of radioactive materials in the United States. Their portion of the CFRs for radiation protection can be found under 49CFR parts 170–175. Specifically, 49CFR173.403 defines levels of activity permitted in different categories of packaging, shipping papers, vehicle placarding, and general safety procedures to be followed in the shipping of radioactive materials. Receipt of radioactive material by an institution is governed by the rules for "Opening and receiving packages" in 10CFR20. Requirements for shipping radioactive material, such as radiopharmaceuticals, radioactive check sources, and radioactive waste, are covered in these sections of 49CFR.

References

1. International Commission on Radiological Protection. 1990 Recommendations of the International Commission on Radiological Protection. ICRP Publication 60. Pergamon Press, New York, 1991.
2. Taylor L. Organization for Radiation Protection, The Operations of the ICRP and NCRP, 1928-1974. DOE/TIC 10124. U.S. Department of Energy, Washington, DC, 1979.
3. Stabin MG. Radiation Protection and Dosimetry. Springer, New York, 2007.

4. U.S. Food and Drug Administration. Guidance for Industry Developing Medical Imaging Drug and Biological Products, Part 1: Conducting Safety Assessments. U.S. FDA, Washington, DC, 2004.
5. U.S. Food and Drug Administration. Guidance for Industry Developing Medical Imaging Drug and Biological Products Part 3: Design, Analysis, and Interpretation of Clinical Studies. U.S. FDA, Washington, DC, 2004.
6. Radioactive Drug Research Committee (RDRC) Program. US Food and Drug Administration. Available at http://www.fda.gov/cder/regulatory/RDRC/default.htm.
7. Atomic Energy Act, Public Law 83-703, as amended, 42 USC 2011 et seq., 1954.
8. Agreement State Program. US Nuclear Regulatory Commission. Available at http://www.nrc.gov/about-nrc/state-tribal/agreement-states.html.
9. Siegel JA. Tracking the origin of the NRC 30-mCi rule. J Nucl Med 41:10N–16N, 2000.
10. Available at http://www.nrc.gov/reading-rm/doc-collections/cfr/part035/.
11. Rutar FJ, Augustine SC, Colcher D, Siegel JA, Jacobson DA, Tempero MA, Dukat VA, Hohenstein MA, Gobar LS, Vose JM. Outpatient treatment with ^{131}I-anti-B1 antibody: radiation exposure to family members. J Nucl Med 42:907–915, 2001.
12. Marcus CS, Siegel JA. NRC Absorbed dose reconstruction for family member of ^{131}I therapy patient: case study and commentary. J Nucl Med 45:13N–16N, 2004.
13. Siegel JA, Marcus CS, Sparks RB. Calculating the absorbed dose from radioactive patients: the line-source versus point-source model. J Nucl Med 43:1241–1244, 2002.

Index

Printed in the United States of America.